Vietnam365

OUR TOUR THROUGH HELL

Vietnam365

Our Tour Through Hell

Memoir of
David J. McCormack

Written by Karen Angelucci

Acclaim Press
MORLEY, MISSOURI

Acclaim Press
— Your Next Great Book —

P.O. Box 238
Morley, MO 63767
(573) 472-9800
www.acclaimpress.com

Book & Cover Design: M. Frene Melton
Front cover inset photo courtesy of David J. McCormack.
Cover background photo courtesy of Bennie Koon.

ISBN: 978-1-942613-68-8 | 1-942613-68-7
Library of Congress Control Number: 2017907032

First Printing: 2017
Printed in the United States of America
10 9 8 7 6 5 4 3 2 1

Contents

This book is dedicated to
Vietnam Veterans

Introduction

It's been forty-seven years since David J. McCormack bravely served the country as an Army Ranger in the Vietnam War. I was only three-years-old at the time of his service. My family was no doubt watching *Hee Haw* and *All My Children* on television. My parents protected my sisters and I from the images of war on the nightly news. Vietnam was too current to teach in schools. Society was listening to The Beatles and The Rolling Stones when David and our troops were crossing into Cambodia. Protestors were angry in the streets and college students were tragically shot while objecting the war.

I was surprised that David asked me to write his memoirs. I tend to walk through doors when opened, but had no idea what I was getting into. I'm compelled to share his memories so that others might understand the lingering and profound effect that war has on its soldiers.

I'm sure there are many untold stories and this one is truly out of my comfort zone because of its harrowing content. I've spoken to other Vietnam veterans and they seem to remember the 365 days of their tour but can't recall what they had for lunch the day before.

I can only hope that David finds peace through telling his story. He once told me that our work together has helped him rejoin society. For this alone, I'm glad to curate and share his memories. There must be a reason why I was chosen to hear him and see his tears during our interviews.

As with all my journeys, I meet amazing people. This book will no doubt open my life to more folks and their incredible lives. I want to thank retired SFC E7 Robert W. Griffin for allowing me to use his spectacular photographs to help tell this war story. I also need to acknowledge Robert W.M. Laughlin for his keen editorial eye and willingness to help me put this memoir together. I can't even begin to express my gratitude to all the men and women who serve, and have served, this great country and the sacrifices they make for our freedom.

Gratefully,

Karen Angelucci
Spring 2017, Kentucky

Ranger Creed

Recognizing that I volunteered as a Ranger, fully knowing the hazards of my chosen profession, I will always endeavor to uphold the prestige, honor, and high esprit de corps of the Rangers.

Acknowledging the fact that a Ranger is a more elite Soldier who arrives at the cutting edge of battle by land, sea, or air, I accept the fact that as a Ranger my country expects me to move further, faster and fight harder than any other Soldier.

Never shall I fail my comrades. I will always keep myself mentally alert, physically strong and morally straight and I will shoulder more than my share of the task whatever it may be, one-hundred-percent and then some.

Gallantly will I show the world that I am a specially selected and well-trained Soldier. My courtesy to superior officers, neatness of dress and care of equipment shall set the example for others to follow.

Energetically will I meet the enemies of my country. I shall defeat them on the field of battle, for I am better trained and will fight with all my might. Surrender is not a Ranger word. I will never leave a fallen comrade to fall into the hands of the enemy and under no circumstances will I ever embarrass my country.

Readily will I display the intestinal fortitude required to fight on to the Ranger objective and complete the mission though I be the lone survivor.

Rangers lead the way!

Soldier's Creed

I am an American Soldier.
I am a warrior and a member of a team.
I serve the people of the United States, and live the Army Values.
I will always place the mission first.
I will never accept defeat.
I will never quit.
I will never leave a fallen comrade.
I am disciplined, physically and mentally tough, trained and proficient in my warrior tasks and drills.
I always maintain my arms, my equipment and myself.
I am an expert and I am a professional.
I stand ready to deploy, engage, and destroy, the enemies of the United States of America in close combat.
I am a guardian of freedom and the American way of life.
I am an American Soldier.

Ballad of the Green Beret

BY STAFF SERGEANT BARRY SADLER, © 1966

Fighting soldiers from the sky
Fearless men who jump and die
Men who mean just what they say
The brave men of the Green Beret

Silver wings upon their chest
These are men, America's best
One hundred men will test today
But only three win the Green Beret

Trained to live off nature's land
Trained in combat, hand-to-hand
Men who fight by night and day
Courage peak from the Green Berets

Silver wings upon their chest
These are men, America's best
One hundred men will test today
But only three win the Green Beret

Back at home a young wife waits
Her Green Beret has met his fate
He has died for those oppressed
Leaving her his last request

Put silver wings on my son's chest
Make him one of America's best
He'll be a man they'll test one day
Have him win the Green Beret.

Facts and Statistics

- 3,403,100 (including 514,300 offshore) personnel served in the Southeast Asia Theater (Vietnam, Laos, Cambodia, flight crews based in Thailand, and sailors in adjacent South China sea waters).
- 9,087,000 military personnel served on active duty during the official Vietnam era (August 5, 1964-May 7, 1975).
- 7,484 American women served in Vietnam. 6,250 were nurses.
- Wounded in action: 303,704
- Hostile deaths: 47,378
- Severely disabled: 75,000--23,214 100% disabled; 5,283 lost limbs; 1,081 sustained multiple amputations.
- Married men killed: 17,539
- Peak troop strength in Vietnam: 543,482 (April 30, 1969)
- National Guard: 6,140 served; 101 died
- Last man drafted: June 30, 1973
- 97% of Vietnam veterans were honorably discharged
- 240 men were awarded the Medal of Honor during the Vietnam era.

Deaths	Number	Average Age
Total	58,148*	23.11 years
Enlisted	50,274	22.37 years
Officers	6,598	28.43 years
Warrants	1,276	24.73 years
E1	525	20.34 years
USMC 0351	1,122	20.46 years
11B MOS	18,465	22.55 years

**38,224 were in the Army*

Source: National Archives, www.archives.gov

Vietnam365

OUR TOUR THROUGH HELL

Memoir of

David J. McCormack

CHAPTER ONE

The World is Watching!

June 14, 1970

We arrived in Binh Hoi at 10:00 am. The Pan Am Boeing 707 came in from the east, taking a steep dive to avoid enemy fire. The plane hit the runway with tires squealing. As the engines reversed thrust, the pilot announced, "Disembark quickly, gentlemen!"

All of us – 217 men – ran down the steps. I had been told that, upon arrival, some troops would lob a grenade or two just to scare the newbies. I thought I was prepared for anything. Upon landing, there were already real explosions outside the Binh Hoi perimeter. Just when I was thinking, "This is the real thing," fifty feet from my right, *BOOM!*, another explosion went off.

A specialist second class E-5 grabbed me by the shirt collar and waistband of my khaki uniform. After slapping me on each side of my head he said, "I got 89 days to wake up and a newbie is not going to get me killed!" After putting me back on the ground he continued, "I've got a wife and kids back in the World."

We were directed to change out of our khaki dress uniform into jungle fatigues, and then loaded onto busses and headed for the 90th Replacement (Battalion) at Long Binh. Ten of us with the 5th Infantry of the 2nd Battalion were bussed next door to II Field Force Base Camp. This is where a Spec-5 informed me where the Headquarters and Headquarters Company was located. We were told two trucks would pick us up in about two hours. I said, "What do you mean, I was told that Head and Head was here on this base camp."

"Not since mid-'68, trooper, they moved it to the high plane," the Spec-5 answered.

About two hours later a truck pulled in. "All troops for Head and Head and 5th of the Second climb on board of my deuce and a half." The sergeant driving had on web gear, helmet, flack vest and an M16 in hand. I thought, "Oh shit, what am I into now?"

As the ten of us climbed in the rear of the truck, one man in the back dressed and equipped the same stated, "Welcome to the real war, troopers!"

Right out of the gate we turned east on Highway 1. After about two minutes, we came upon a village and the E5 with us said, "Welcome to Honi Village, gentlemen, the Viet Cong from Honi will welcome you in their own way, in time."

At the end of the rather large village, more like a town, we turned right, going south. After about 6 to 8 miles we pulled up to a gate. A large sign, 10 to 12 feet tall and 14 to 16 feet wide said, "GOD HELP US!" with an image of a Bengal tiger on the right looking straight at us, and a # Field Flash to the left. The top row of lettering read, **"H & H 5/2."** Directly under it read, **"30th ARVN Rangers."** Under that read, **"5th Special Forces MACV Advisory Force."**

To the right of the sign there was a body of a Viet Cong. The corpse, maybe three days' dead, was hanging from a hangman's noose. I would later learn about the thirteen loops in the noose and how the Buddhist's believed if a person was hung that way they were Hell bound. One of the newbies inquired about the dead Viet Cong and the reply was, "Hey man, the Tigers don't mess around! They mean business."

At the same time, I scanned the troops inside of the gate. I observed four or five Green Berets. I also saw a man in solid black dress. "My God," I thought, "That is a Navy Seal, just like in the movie *The Green Berets*, starring John Wayne!" There were two men wearing black berets, and I would soon learn they were with Military Assistance Command Republic of Vietnam Advisory Force or MACV.

After being familiar with only stateside duty, I thought I would surely die and now I was in Hell.

The guard opened the gate and let us enter. There were only three buildings on this little 100 x 100 yard plot of land. In the next 14 days, I would especially learn one building, the ARVN officers club, and another large building called the hooch for all the Special Forces and Special Operation Forces. The third building was the Headquarters, and of course there was a privy. We went through another gate to enter the Head and Head on the south side…100 x 200 yards.

There were a number of small buildings centered on the north end of the compound. We made a large horseshoe U-turn on the sand berm road and stopped at a small building close to the north and west wall.

An E-5 buck sergeant greeted us with, "Welcome to the war, gentlemen!"

I made a mental note while making the loop that there were two M42A1 Dusters respectfully in the southeast corner and the southwest corner. Each corner was sand-bagged and cutouts had been created in these corners. One Duster stood in reserve on the east wall.

Two M55, quad mounted .50 caliber air defense systems were present in the center on the compound. I would observe later that the "Tops Pop Corn Machine" (as it was called) would be normal on the compound, but only the top two guns mounted were working. The gun truck would pull duty from 1900 hours to 0700 hours most nights. The three-man crew were all profiles, still having wounds in the healing process and all having been awarded the Purple Heart — a norm and common award in the 5th of the 2nd, Delta 71st and I-29th.

Men moved about, some armed, some not. The ten of us were put in a small hooch that had nothing but some old canvas cots, with one wool blanket and one pillow per cot. There was no electricity; no frills; the walls from the floor and up forty-two inches were very similar to railroad ties and open except for mosquito netting on the upper portion of the walls.

The sergeant in charge told us to stay close to the hooch, as we were in a dangerous place. He said he didn't want newbies getting hurt in some accident. "Just hang close to the transit hooch and try to relax," he said.

I pulled out my copy of Walter J. Black's, Zane Grey, western novel, *The Vanishing of America*, and tried to read a bit. That novel and my French-English dictionary were the only books I brought with me. However earlier that day, three religious books were given out in the rush. The Jewish received a Hebrew Bible, the Catholics received a military Missal and the New Testament for Christians, which I took.

The sergeant came for us to have our first meal on the compound. We walked a few feet apart to go to the mess hall. We entered what would resemble, for the non-military civilians, the 1960s cattle bar, with shipping pallets down for the floor. The sergeant directed us, "Over there we have steak and potatoes. Over there is the salad, and desserts are over there on the counter."

I wasn't even hungry; my body was on overdrive. I didn't need food. After twenty-four hours on that 707, I thought "How is a man supposed to eat after sitting in the same seat that long?"

We went through the chow line. The tables were picnic style, with poles for legs and rough sawed lumber for table tops and seats. Upon inspecting them, we sat while trying not to catch a splinter.

I sat down and observed a thousand-legged worm crawling down my spaghetti. I took a spoon and flicked it at one of the local women cooking in the dirt floor cooking area. It hit her in the upper back, causing her to have a cursing fit in her native language. I did not eat anything at that meal. I just put the tray in the window and went back to the transit hooch. About sundown I needed to go to the latrine. I asked the sergeant where it was. He said pointing, "Just follow the chain link fence there about seventy-five yards. It's next to the arriving gate." I took off, thinking "easy enough."

As I walked the fence line toward the latrine, the smell that permeated through the 102° Fahrenheit degree temperatures and 100 percent humidity of my new home thickened and was terribly indescribable. I pushed through the screen door and there was one rough welded urinal. It was rusty, with two urinal cakes laying in the bottom. There were three stalls with no doors.

The roof to the privy came down low and was lit by one little 25-watt light bulb. It was all but dark inside and growing dark outside. I walked into one of the stalls. There, a 12-inch gas pipe came up out of the rough poured floor. The shape of the toilet seat was torched out of a 1-inch plate still in the tough torch cut. The edges were still sharp and ragged. The seat was only six to nine inches above the floor, making it difficult for a six-foot man to straddle his feet, all while trying to keep trousers from touching the muddy floor.

This squat was made possible by cupping the family jewels with one hand, and boosting the body up with the other. On a good note, there was soft toilet paper hanging from a peg on the wall. Thankfully, I had my Zippo lighter to assist me. I finished and made my way back to the others.

At 1830 hours, I watched and noted a 21-man guard mount inspection by the officer of the day. Three-man fire teams were posted in six bunkers and the watch tower. I was in my bunk by 2200 hours that night.

I awoke at sunrise, got up and just hung around the hooch. I lay down again on my cot around ten hundred hours when heavy incoming artillery hit the southwest end of the compound. Then it stopped. The other nine newbies in the hooch were running in all directions, not knowing what to do or where to run. The sergeant chased everyone down and informed us that when this happens to get into one of the air raid shelters. Within moments we saw an ambulance taking four Duster gunnery crewmembers headed for the main hospital in Long Binh.

The following day, all ten of us newbies were loaded on the same deuce and a half we came in on. We were taken back to the II Field Base Camp. Unlike the arrival route, we took a military road rather than the local road. We were taken to the post exchange to buy shaving cream and whatever we needed. Understand that we weren't allowed to bring any aerosol cans over on the flight for fear of explosion when the cabin depressurized. During this ride, about two miles out, we observed the ruins of an old French fort that fell in 1954.

We arrived at the II Field Base Camp, and a fine PX it was. We purchased our shaving cream and any little thing that we could use to make life more comfortable on the compound. I was so involved looking around that I didn't notice the novice buck sergeant had left. He did not do a head count and I was left behind. I looked out the PX door and the other men were gone. I walked out of the PX very shaken.

I sat down on a short wall, thinking about what I should do, when a captain walked up, looked at me and said, "Don't you stand and salute in your Army?" I could only sit and stare at him. He then said, "I guess you've been here too long already," and he walked away.

I gathered my thoughts of getting an Article 15 and thought that it really wasn't that far from the Duster Compound — surely I could get a ride. I walked about 200 yards to the highway and approached the military police checkpoint. The MP on duty asked, "Where are you going, trooper?" Not knowing where for sure, I said, "Hey man, everything is cool."

I started walking east and flagged down a local man in a Lumbretta three-wheeled scooter. Sitting in the back of the scooter were two old women who were sitting opposite of each other. I asked the man if he knew where the Duster Compound was. He nodded that he did. I showed him a twenty-dollar military payment certificate. He motioned for me to get in the back on the floor between the two old women. We rode a short distance and one of the old women shouted in her native language.

Ahead of us was a provincial police officer directing traffic with total and absolute power of arrest on anyone. The old women both pushed my head down, as I was sitting cross legged in the floor. They both pulled their long skirts over my whole body to hide me from the police officer. The officer stopped the little three-wheeled machine for about two minutes. I could vaguely see him through the somewhat sheer skirts of the two women.

I turned and looked straight ahead and thought, "My God, these people don't wear underwear!" We went through the village after police waved us on. When I discovered the women were naked under their dresses, I closed my eyes and hid my eyes with my hands.

After we cleared the village, they uncovered me. All three of them were laughing at me as I removed my hands from my modest eyes.

We finally pulled up to the special operations gate where we had entered the day before. The captain of the A Team was watching as I exited the rear of the small machine. As I handed the old man the $20 military payment certificate, the captain shouted, "Hey, you never give the locals an MPC, that is more than he will make in a year farming!" The captain continued to shout at me. I could feel his breath on me, "Never get in a vehicle with a local again! You are very lucky to be alive. You can get your throat slit doing that."

The captain continued on me after he handed the local man something folded up, "Come on in, I could have you Article 15'd for that, but you are a newbie and I'll let one go this time." Just when I thought he was finished he said, "Never do anything like that again, understood?"

To that I replied, "Sir, yes sir!" as I hustled through the two gates back to the transit hooch.

At that point the novice sergeant caught up to me stating, "The colonel sent me back looking for you, telling me if anything happened to you I would face general court-martial and spend time in Leavenworth Prison with a dishonorable discharge." He took a deep breath and looked me harder in the eyes, "Never do that again, you could be killed!"

I replied, "Okay, okay!"

Later, Chief Silvers sent a couple of enlisted men for me and moved me into the Battalion Recovery and Deployment hooch with 12 men of the platoon. I was moved in with a roomie that only had a few days left in the country and had been wounded in the 25th Infantry.

The wounds on his back was mostly healed and he was being reassigned to Headquarters and Headquarters. Due to the wounds to his back muscles, he could no longer carry the heavy loads required to be a grunt. He told me when he left in a few days he would see that I got his poncho liner. There was a shortage of these on the compound. He also said I could have his footlocker, which also turned out to be a coveted item due to shortage as well. I gladly accepted any items he wished to leave me to make my stay more tolerable.

About dusk my new roomie came to me and said, "I'm going into the village to see my girlfriend. If anyone asks, cover for me, say I went to the latrine." After I agreed to cover for him, he pulled out a sawed off single shot 12-gauge shotgun and a sack of shells and tucked his KA-BAR combat knife under his belt buckle. He looked at me, winked, and said, "I'll pick up a few grenades from the chief's bunker, and be back in two or three hours or so."

He walked out a path that was well worn through the minefields in front of the M42 Duster in the southeast corner of the compound. Sometime later, early in the morning hours, we were awakened by several grenades and distant shots.

When my roomie returned, he was out of breath and wet from sweat. "The bitch sold me out to the VC and I came real close to being had," he whispered loudly, gasping between the words.

The watch had changed on the Duster and he said, "A kid near shot me as I was coming back in. That will be my last trip into the village!"

The next morning a staff sergeant rounded up a number of us new troops to try our hand at driving a 2 ½ ton truck. Our duty was to drive to the Saigon River and haul back sand to put down on the berm road and inner roads to keep the mud down.

When asked about my experience with trucks, I replied that my license was only for jeeps and light trucks. He replied, "Let me see your license." After checking my license, he handed it back to me and said, "You have a 2 ½ ton truck now. Go saddle up in that deuce and a half."

I quickly replied, "Not without a rifle!'"

CHAPTER TWO

Chief Silvers

The first Monday I was there after the sand haul, a couple of men in the platoon wanted to take me out to the Special Forces-Special Operations side for an initiation. I had heard of these stunts stateside and knew what to expect, having grown up close to Fort Knox, Ky.

We went to the front porch of the officer's club, where one of the E-5 sergeants was sitting with two Tiger Rangers. He told me that they were assigned as his bodyguards. The sergeant was just as excited as I was to be around the Tiger Rangers.

He then told the two bodyguards to open their shirts. The first guard pulled his shirt open and showed me a scar from a 7.62 Russian rifle. "This is where the Viet Cong shot me," he said with great pride. The Ranger then told a fact, he said, "Whatever happens to one of the advisors, our colonel does back to the bodyguard." He explained while pointing to his shoulder, "See, when I took this bullet, the colonel shot both of the bodyguards in the same shoulder and we all went to the hospital together. We stayed together and they put one guard on both sides of me. We were all released together. If an advisor gets killed, the colonel kills the bodyguard," he added.

The sergeant told the other Tiger Ranger to pull his blouse full open. He had a scar on the left side of his lower ribcage. The sergeant said, "This Tiger jumped in front of me… took a bullet for me!" With intensity, he said," I smoked that VC with my battle rifle on full auto."

While we all listened to these valor acts, the sergeant reached into his wallet and pulled out some piasters, which was the Vietnamese currency. He told one of the bodyguards to step inside and buy a brand 33 beer. The bodyguard delivered my beer and after I tasted the strong rice flavor, I said, "Wow, this has some kick and somewhat strong!" The sergeant smiled and replied, "Yes, and really good!"

About that time the sergeant held his clenched fist to his heart. The bodyguards all but panicked until the man said, "Sometimes I have a little heart trouble." The sergeant laughed and the men knew he was just teasing.

The sergeant then pulled out an M26 hand grenade from his jacket, the kind that looks like a pineapple. "You know if a man is right handed, he will hold the grenade in the right-hand with charging handle up, pull the pin and throw." He puts the grenade in the other hand and said, "But if a man is left handed, he can turn the top down, pull the pin, and throw it." The sergeant then said that because of a left shoulder injury, he has trouble with fumbling things. He said, "Hold the charging handle like this," at that time the grenade fell from his hand. He shouted, "Oh My God!" as it hit the floor and, of course, the handle flew off.

I stood about one second as the sergeant and bodyguards dove into the door of the club, and my two platoon members ran.

I stepped around the corner and counted off 30 seconds, knowing it would have been a five second fuse. The grenade wasn't cooking, so I knew it was a dud or a nonworking grenade that had been disarmed.

After 30 seconds, I stepped back around the corner, calmly picked up the pineapple grenade and charged the handle. I handed it to the sergeant and said, "You need this back for next time."

The sergeant looked at me and said, "I have seen newbies dive on it with their body and run every which way, but you are the calmest, coolest man I've ever seen pull a stunt on me. I like you!"

The next day at 1200 hours, I reported to the sergeant and the captain of the A team for training.

On that first day of training, a warrant officer who would be leaving in a few days came for me saying, "Man, you need to report to the new chief as my replacement, and you had better hustle. You report like you have never reported before."

I went down to Battalion Deployment and Recovery and reported to Russ, the chief's clerk and radio operator. "The chief will see you now," he said.

I had been told to hold that salute until it was returned. I walked into Chief Silvers' bunkered office, came to attention, saluted and held the salute. I could see he was looking over my 201 file. On his desk, he had spread out my two Dana certificates on Diesel engines, one for completion on gasoline repair and one for rebuild. He also had my Chevrolet Master Technician certificate I had earned before Basics, lying on top of the file.

He stood up, saluted, released the salute, then sat back down without speaking.

He picked up a letter of commendation signed by William J. Furlong, the company commander of the 603rd Transportation at Fort Polk, then went through the riot control training I had, as the 603rd doubled as a riot control unit. On a few Thursdays, all Fridays and some Saturdays, we had been trained by two Green Berets and two Navy Seals.

He then stated, "I think you are worth saving." He looked at me in the eye and said, "I knew your father from Fort Knox Transportation Motor Pool. I know all about his war record with the 24th Infantry in the South Pacific. The man is no doubt my hero." Just as I was feeling overjoyed with pride he said, "However, that will have no bearing on his son."

"I could put you on as a Duster crewmember, but within 90 days you will be dead or going home on a hospital flight with body parts missing. You will be a total basket case, most likely living in a wheelchair for what life you have left with little quality of life."

After a slight pause, he asked, "Do you understand me, specialist?"

"Sir, yes sir!" I firmly answered.

"I am going to give you sixty seconds to make a decision — work for me no questions asked, or go on a Duster crew." Before my minute was up he added, "If you ever question my orders, I will that instant make you a Duster crewmember, and I have already given you the true expectations of what will become of you."

The chief unbuckled his watch propped it up on his desk and pointed to it, "Time starts now."

I did not have to think, I would work for him, whatever I had to do to serve him and all the Battalion and my country as so many in my family before me.

The chief stated, "Time's up. What is your answer?"

I replied, "Sir, I will work for you, no questions ever, I am here to serve you and my country, sir."

The chief replied, "Okay, see that black Jeep out there?"

I replied, "Sir, yes sir."

"That Jeep is my Jeep, I am assigning you with the task of bringing that machine up to 100 percent. You will put five new recap tires and tubes on her to start. You can only patch a tube one time, and the interior of the tire one time. Any damage more than a small nail hole, you must replace it and take the tires from it up to the searchlight team for their Jeeps."

He continued, "Never put a tire on with a Vulcan patch or repair on my Jeep as it can create heat and cause a failure. It needs plugs, points, a complete going over. After the tires, do whatever is necessary."

I stood there taking it all in, feeling excited and up for the duty, when the chief then said, "You can do anything you desire, but if it ever breaks down with me in it I will help you repair it. Then we will return here to the Duster Compound, where I will take you out back with the retrogrades, take off my bars, beat the hell out of you, and then assign you to a Duster and you know what that will mean."

"Of course," I answered, "Sir, yes sir!"

I followed the chief towards the tool room. "The local's name is Tran Van Loc, we call him Charlie. He will assign you a toolbox and help you get you started with those tires. You can trust him; his father is a government official in Honi Village."

"You will start now and work until 11:45, at which time Cookie will bring you a meal to eat here in motor service. At 1200 hours you will report to the E-5 sergeant on the Special Forces side of the chain link fence for training under the supervision of the captain of the A Team."

I stood there accepting every command. "You will be exempt from any other duty for ten days." The chief handed me the key to the padlock on the tool room." At that moment, I was proud and eager to fulfill the chief's confidence in me.

I continued to stand at attention, holding keys to my military future. "The Charge of Quarters, the CQ, will wake you at 0430 hours each morning. You will report here at 0500 hours to work on my Jeep each day until it is 100 percent. At 1800 hours each evening Cookie will bring you the evening meal and someone of my platoon will come for the tray. Do you understand, specialist?" asked the chief.

I answered, "Sir, crystal clear, sir."

"Charlie will assign you the best Echelon 1-2 toolbox he has. He will get you started on the tires and help you do these things, the way they must be done here in the war."

We approached the tool room door, a Dutch-style door with the upper door open.

"Charlie, this is Specialist McCormack, assign him the best toolbox you have and take the jack and tools, help him and show him the right way to mount five new tires."

Charlie replied, "Yes chief, sir!"

Charlie handed me my assigned small toolbox, pulled out a mattock with the spike cut off, a small barrel jack and two tire irons.

Charlie and I began the process of installing the new tires. He demonstrated placing the mattock next to the tire bead and using an 8-pound sledgehammer, and how to use the tire irons to remove and replace the tires and tubes. We inspected the rims for gouges and anything that could damage a tube. Charlie used a file to smooth these places. We worked as a team.

We talked. He told me about his father's position in the local government and how that kept him out of the military. He had been with Chief Silvers and the unit since the summer of 1966.

I noticed a 90cc Honda, and he said it was his and he had bought it with his salary. I would learn in time, after seeing Silvers close the door repeatedly to the office, that Charlie was bringing information in about the Viet Cong. Such information was passed on to MACV Rangers and the captain of the A Team. No one else on Duster was allowed to know about Charlie bringing intelligence in or the Viet Cong would have killed Tran Van Loc immediately.

After the tires were finished, I pulled the worn out spark plugs. I had to use 5-ton gas spark plugs using some extra plug washers to shorten the reach. It was good though; the heat range was higher and hotter and would burn cleaner, with cleaner combustion in the cylinders.

My next step was to replace the points and condenser. I took the volt-ohm meter and went over the complete ignition to make sure it was 100 percent. It was.

Now to set the ignition timing.... I took to heart what the chief said, "Do anything you want to the Jeep." So, I thought, these military vehicles are literally "de-tuned."

I took a string and measured around the crankshaft pulley. I used a tape measure and calculated eighteen degrees before top dead center. That was the factor specification for the '61 Corvette fuel injection that was what I used on my '57 Chevy I called the "Grey Ghost"; it ran best changing gears at full power shift at 7,200 revolutions back on the street.

I used the corner of a file and cut a new notch at 18 degrees before top center; a touch of white paint and with a timing light set the timing.

Now for the valve lash on the mechanical value tappets.... Our instructor back home, Elvin E. Smith, who was the best in the nation in my opinion, had taught us students to make these adjustments on a still engine using a process of running mates or sometimes called companion cylinders.

I also used the Corvette hot rod specifications used on the old '57 to '61 fuel injection camshaft part number 3736097. Hot-rodders were pulling the tappets down from twelve thousandths on the intake values to eight thousandths cold. This would come out to six thousandths and a hot engine. The exhaust value factory specifications were eighteen thousandths hot and we hot-rodders were setting them cold at sixteen thousandths, that would come out to fourteen thousandths hot.

Changing from the twenty some odd thousandths of military specifications down to what I had just done would open the values much wider, and hold the values open much longer in degrees on cam shaft duration.

I put the value cover back on and started the little machine up. She had that loping sound so sweet of a three-quarter racing cam in a small block V8 Chevy.

Now I needed to dump a serious fuel mix into the cylinders. I walked about fifty feet to the Battalion welder's shop. I asked to borrow a small hand drill to clean the welding and cutting torch tips. This tool is what hot-rodders were using to bore carburetor sets up to ten thousandths oversize. In those days I had memorized each bit size and could take from memory the steps on the bits. The end result was after removing the main metering set from the carburetor, I bored the set from between twenty to thirty thousandths oversize.

The chief approached me when I told him the Jeep was finished. He advised me to rob and install a seat belt from one of the retrograde Dusters. Then he worked with me to rob a rifle mount from the floor of the Duster and to cut down another seat belt to mount my AR-15 vertical next to the door opening, so I could hit a quick release as if I had to bail out of the vehicle.

At the end of the fourth morning, the Jeep was finished.

The day after completion of the Jeep, the chief put me on cleaning up a mess in the mini-can. It needed to be cleaned and I also needed to clean the serviceable magazines for use, to put thirty, 20-round magazines into service in bandoliers and another 600 rounds in bandoliers.

He took me to the Deployment and Recovery red status bunker and said, "The rest of the troopers who have E-5 status and below normally do not handle M56 grenades unless we go on red status, which means we are under siege. However, this case of grenades is yours.

"Now, go get one of the ammo bags that came with the Dusters and you need to carry six grenades with you on all missions. Chief handed me a metal olive drab box, "Here is your case of grenades for the personal M79 launcher that I will acquire from the ARVN Rangers. I will also acquire an AR-15 for you. Neither will be registered with Head and Head."

Chief Silvers continued with my particular training, "The Colt M16 you have been issued is for guard mount and show. After guard mount, go back to the hooch and switch off to the AR-15."

I stayed at attention, "Tomorrow morning we will take a business trip to Long Binh. We will acquire a large wall locker for you, the double door with drawer in the bottom to lock all your belongings and the three weapons.

"No one will mess with you, no one gives orders to you, only me. You work for me alone. When you have greenline, the officer of the day, of course, will be in charge; however, I will be watching every move when it comes to the men in my platoon. The places you will be traveling to, well those officers, remind them I am your commander in charge and you answer to me alone. I have this agreement with the Battalion commander, the sergeant major and the major."

After a short pause, chief asked, "Do you understand me, specialist?"

"Sir, yes sir," I replied.

About this time Chief Silvers told me I must take two blouses and with a razor blade remove the U.S. Army and my name. When it came to my dog tags, he said to put one through the laces of a boot. He then told me to always take the chain and one remaining dog tag and put it in my pocket. By doing so, it would appear to the enemy that I was one of the 2,000 civilian contractors. Normally the enemy would not bother with such as they spent so much money in the country.

We took the Jeep and went to a large ordinance shop in Long Binh. The chief approached a major. He gave him a drawing of trigger bars that he desperately needed for the Quad 50 back at the compound. Chief made it clear that he needed them as soon as possible.

"Well chief, it will be about six weeks," the major said.

The chief went off, "What the hell do you mean six weeks! I've got people dying out there! Well, I see you have one of my 'Little Joe' engines there! What are you doing, souping it up for the go cart race to impress the Donut Dollies!"

Barely taking a breath, the chief added, "You know what I need. I need someone to cover my man's ass out there with the 79 on his road trips. I might need your ass riding shotgun! I want you to talk to your bird right here and now. If I don't get results, I will speak to a star!"

"Okay, okay! Chief give me a few days," pleaded the major.

It only took six days before I would return to Long Binh for a large box of trigger bars.

CHAPTER THREE

Bitch in Heat

In the coming days I would report for special training. I reported to the Military Assistant Command, Vietnam (MACV) to take orders from an E-5 and the A Team commander. The sergeant looked me square in the eyes and said, "Welcome to my training school!"

My first lesson covered Air Assault Heliborne. There was a tower on their end of the base. It was twenty-five feet high with an old Huey UHIC helicopter fiberglass hull with the main airframe sitting on top of the tower. There were steps going up to it. It doubled as a watchtower during the hours of darkness.

The sergeant brought me what was basically a parachute harness modified sort of "D" ring. It was mounted on the front somewhere about armpit level or so. The strap going across the front with the "D" ring was heavy and somewhat wide.

The sergeant assisted me in harnessing up. He explained what was going to take place, while one of his bodyguards went up into the tower to operate the winch and to assist me. He then hooked me up. I was then winched up and made it inside the old helicopter.

Then the bodyguard gave me the long heavy gloves and, after putting them on, the sergeant talked me through rappelling down after wrapping the rope twice through the "D" ring. That whole day until about 2200 hours was practice and re-practice. The next day beginning at 1200 hours until 1400 hours was repeated practice.

Then a working Huey flew in. He hovered at about 35 feet. At that point the rope was dropped. I hooked up and was lifted up. There were no guns on this Huey, only a pilot, co-pilot and the crew chief, well that was the entire crew.

The crew chief inspected my work to make sure I had properly placed the rope in the "D" ring and I rappelled down and unhooked, and the Huey flew away.

The sergeant said, "You have just earned the Heliborne wings."

I did not really understand the true meaning of that statement at that instant, and did not want to appear ignorant, so I said nothing.

The sergeant then said, "Now it's time for knife skills."

I would learn many combat maneuvers with my knife, as well as how to live from God's land. There were a few plants that were also native to Kentucky. There was a fifty foot by fifty foot natural area, well protected by a lieutenant colonel. He was, what I believe to be, with the 30th Rangers of Saigon. He rarely allowed anyone to touch his little piece of Vietnam. He did allow his plot to be used for training of only one American Ranger at a time, but only if the man was in a platoon attached to his battalion.

One plant was much like the mayapple. A MACV advisor took out his KA-BAR combat knife and cut it from the vine. He split it in half. The advisor, his two bodyguards, and I each ate a quarter of the apple.

Another plant was a root somewhat similar to icicle radish. It is a little like a very small sweet potato. The advisor, with his bodyguards on each side of him, thrust his bayonet into the earth about one inch to the side of the root and with a downwards thrust popped the root out of the earth. Using the KA-BAR, he cut off the tops about one half inch into the root, explaining the tops are deadly poisonous and any man would die a horrible death after eating them. He peeled the outer skin end to end, much like one would a carrot. He handed it to me, advising me that clean water in the bush is ever so precious. He said to pour about two tablespoons on the inner lower corner of my blouse. Wipe it clean, then eat it. Each of us did so. The captain of the A Team with clipboard just smiled as he was checking off items on the training schedule.

I was shown an edible grass to eat. It had zero nutritional value, but would fill the stomach and digest well enough. I was introduced to a native berry that grew wild.

After lessons about living off the land, I learned how to eat meat without cooking it, since fires were not allowed in war zones. The sergeant, along with his bodyguards, demonstrated how to whittle barbs on the small end of a piece of bamboo. This would be used to gig fish, frogs or whatever I could manage to get down my throat. They said anything in the waterway was safe.

We discussed snake meat; however, I was warned that I now lived in a jungle with the two most deadly snakes in the world, the Cobra and the colorful Banded Krait. The latter also called the "two step" or "cigarette snake" by soldiers. When I asked why the nicknames, the sergeant said because you have two steps before you die and/or one cigarette.

The bodyguards demonstrated the sergeant's commands on how to climb up a tree, and, jump from a low limb to kill a deer. One guard dropped in the straddle position to jump on the decoy deer. Then he demonstrated how to hold it in a chokehold and thrust a knife into its throat. Deer were somewhat common in III Corps, which is the region from south of Cam Ranh Bay to north Saigon. In order to pull this off you must instantly cut the throat, bleed him out, and cut away enough meat to eat raw from a hindquarter. Then one must drag the carcass into heavy grass and into the triple canopy and put some distance between himself and the carcass. I have always found deer to be quite beautiful, especially their antlers.

Moving out of the training area to return to the Duster Compound, the sergeant asked, "I know you did some fishing back in the World, do you have tackle back in Kentucky?"

"Oh yes, an old J.C. Higgins rod and open reel with a pistol grip handle. My dad gave it to me. He bought it for himself when mom and dad were going together back about '47."

"What kind of line on that reel?" the sergeant asked.

"Well it's the heavy testline for Kentucky catfish, it's the old black, maybe cotton, whatever was current in '57. That's when dad bought a little lone star 14-footer with a 10 HP Evinrude," I answered.

"Very good, write your mom tonight, and have your dad pull off twenty-five feet of that line with two eagle claw hooks. Ask her to put them in a letter ASAP."

"You can cut a piece of bamboo and build a cane pole like you did back home. Use a bug or whatever for bait. You must be and have total concealment; only the end of the pole coming out of the concealment. It would be much better after dark, especially dark of the moon. Tie a small stone for the weight; it will not take much for a float. Find a small stick and break it. Do not cut a clean edge, about the size of your fore-finger. Tie just a single loop just like the first step when you tie your boots."

The sergeant and his bodyguards then wrapped up our session and the sergeant looked at me and said, "Got all of that, any questions?"

I replied, "No questions, I got it."

In a return letter, Mom said dear old dad made this statement at the line and fishhook request, "Our boy is not in a war, he is on vacation."

"Remember this," the chief said, "Viet Cong recon teams many times are two-man teams, spread out some distance apart; it is best to be up a tree. Now let's assume you are in the bush up a large tree, a number of Viet Cong are in a group and do not move on. You need to be 20 feet or more up the larger tree. Find an opening where a grenade will not be deflected, pull the pin and release the charging handle. Cook it for 1.5 seconds and drop it straight down. At that instant, move around and use the main trunk to shield yourself from the blast. Drop as many as necessary within reason. You want to know how to break out if you are surrounded by a number of Viet Cong?"

I stayed at attention and took it all in. "Well, I would pull out the M79 grenade launcher, about 12 fragmentation M56's, and one pouch of 400 rounds for the AR-15." Sergeant got down on the ground to demonstrate the maneuver. "Now lay down prone on your stomach, fire only semi-auto around 90 degrees of the 360-degree circle." While rolling, the sergeant continued, "Now roll over on your back, pop up and fire 90 degrees opposite the first field of fire. Then you will want to drop down flat of your back, turn 90 degrees and repeat what you have just done."

"Now choose the weakest point and launch six to nine grenades with the M79. You want to charge that weak spot and prepare to use the AR-15 and M56 grenades as necessary. When you have expended all 12 grenades through the M79 field strip, throw the three components in three different directions so the Viet Cong most likely will not find all three components and use it against us." Standing up, the sergeant then said, "Now run! Double time in a zig-zag manner; that, you will learn, is how the VC does it."

Walking over to a table loaded with assorted items, he said, "Now it is time to learn how to disassemble and short fuse an M56 fragmentation grenade." He picked up an olive drab grenade out of its carton. It had a yellow band around the top. There were numbers and markings stamped in yellow as well on the body.

"I'm not going into a great deal of detail on this, however, with some of the tape to construct a 'tower' so to speak after cutting out the fuse and filling the tower with gun powder from the 5.56 cartridges using a short piece of hacksaw blade, the grenade will detonate instantly when the handle is released."

The sergeant laid down the grenade, "Now the c-ration can, such as the beanie weenie can, will hold the handle intact after placing the grenade in the can. Then you will want to take a trip wire, or in this case heavy clear nylon string, and tie off on one end. The other is tied to the grenade.

The sergeant demonstrated how to pierce two small holes with the bayonet opposite each other in the bottom of the can. Then using lacing wire, he tied it to a small tree close to the ground. He showed how to conceal the clear string with loose vegetation. The Viet Cong teams that are tracking me would trip on the string, pulling the grenade out. Instantly, all in the upright position within 17 feet will be killed. Those within about 37 feet if not shielded by a tree or whatever will be wounded. "None of them will be capable of tracking me," he said.

We discussed the enemy. Just as we do, they used track and trail dogs to trail Americans and NATO troops (which consisted of four countries with about one division each, the Philippines, Australians, South Koreans and those of Thailand.)

Showing me some diagrams, the sergeant showed me a few very simple tricks to throw off the dogs. He showed me how to take a small amount of pepper from the packets. Sprinkling as you go, walk backward for 20 to 30 feet backtracking. When the dog hits the pepper, it screws up his or her sense of smell and ability to track for two or three days.

Another lesson taught was called "Bitch in heat." If a Ranger can manage to kill a female dog from either side, skin the animal and literally wear the skin on ones back with glands rubbed on the feet; it drives the track and trail dogs nuts and one can even enter an enemy camp with guard dog encircling the enemy camp without the enemy's dogs going on alert or giving warning.

Next in training, the sergeant is tracking me; I am alone in the bush. I must get four hours or so of sleep to function the next day in hopes of being found from the air and or search teams. I have been taught to climb a large tree and go as high as possible. With two wide long web straps, I secure myself with the two straps with battle rifle and possibly the M79 (if I have not already fired dry and disposed of the weapon).

I then secured my weapons with a strap just under my arm pits and the remaining strap around my hips to hold me and my weapons in place for those precious four hours or so of sleep; both the ammo pouches, one hanging on each side clipped out 5.56 battle rifle ammo, and the six M56 grenades, or whatever I still possessed. The 11-inch combat knife was strapped to my right arm on the top web gear, wearing it of course, with only a canteen of water, the bayonet on the left with canteen and the first aid pouch on the right.

At this point on the last evening of special operations training, the sergeant gathered my gear and put all on a scale to weigh, including the AR-15.

The gear consisted of one web gear, one battle rifle AR-15, one ammo pouch with magazine, one ammo pouch with six M56 grenades, a long roll of lacing wire sealed up, one can of beanie weenies with one plastic spoon held on by two large rubber bands secured around the can, one red smoke grenade, two web straps (that would be wrapped around the poncho roll), smaller pliers with wire cutters, two tent ropes from tent that would be looped together to form a carry strap or rope length of 100 mile per hour tape wrapped around a small wooden pin, and a number of pepper packets from c-ration meals.

I was advised to have one tiger stripe uniform with boonie hat, which I did acquire from the Honi Village at a later time, camouflage kit, short piece of hacksaw blade via 100 mph tape for a hand hold. I was shown how to roll all of these essential survival tools and how to place the tent ropes for the shoulder carry.

Familiarizing me with such other weapons, they showed me a captured AK-47, Chinese Communist Carbine, the S.K.S. Chinese battle rifle and captured grenades.

The sergeant then walked over to me with the two bodyguards and the A Team commander. All four men of different ranks saluted me as the sergeant stated, "Welcome, you are now one of us; go get them, Tiger Ranger!"

I returned the salute and one thing ran through my mind, "Lord help me, what have I gotten myself into?"

The sergeant then stated, "Army Ranger, you have graduated with honors; time to report back to Chief Silvers."

I replied, "Thank you sergeant, captain sir."

I nodded to the two bodyguards. They smiled back at me.

CHAPTER FOUR

Tiger Ranger

I walked through the gate with the AR-15 in hand, web gear and the poncho roll slung over my shoulder. Chief Silvers met me about 100 yards from the gate. He said, "You have your first greenline tonight, use the Colt M16 for inspection. Be 100 percent sure that both battle rifles are clean and then switch off to the AR-15 for actual greenline in the bunker tonight."

Without even saying hello he says, "You and I have a hard four days starting tomorrow morning. Your training is over with Special Forces and Special Operations. We will be leaving as soon as possible for Cu Chi in the morning. For the next four days it will be just you and me, I will work with you and lecture you as much as possible as we drive."

"Get a shower and shave, it's now 1700 hours, guard mount is 1830 hours in the center of the compound next to the flag pole." He finished with, "Just do your best; that is all I want from my platoon."

"Sir, yes sir."

Twenty-two of us enlisted men were present for the guard mount, and it was over in just a few minutes. This is where I learned what a supernumerary was. The best trooper would remain with the officer of the day and sleep in the orderly room. He would only be assigned to a greenline bunker if someone became sick or needed to be replaced for whatever reason.

When released from guard mount, the officer of the day informed us, we must go to the underground war room. In single file, one of the troopers said, "Must be something bad, I've never had to do this before."

We went down sandbag steps, passing wiring that was exposed. The wiring turned out to be high explosive charges set in the sandbag walls of the steps. In case the base was overrun, those charges and two more from tunnels coming from the NCO quarters and officer's quarters could be blown. Everyone that could make it underground would be sealed off until help arrived from gunships, or Puff the Magic Dragon as we called it. The problem was that the aviation unit was 15 to 17 minutes away on a good day.

The CIA had just arrived in two M113's (armored track vehicles). The war room was a little more than 30 feet wide and maybe 50 feet long. The ceiling was burned and finished plywood, as well as all the walls. The floor was concrete, and there were about 40 folding up chairs in rows.

There was a very large map on the wall that had pins with six different color pins in different locations in III Corps and IV Corps.

These indicated the locations of Dusters, Quad 50s and the searchlights, and were moved frequently as needed where there were large buildups of enemy troops.

There were blown up photos lining an entire wall. There were three rows of photos of the same shot taken from the 75-foot tower with a military camera with telephoto lens, very powerful images taken every Sunday at exactly 1400 hours.

These photos were respectfully, current, one week old, and two weeks old. A sergeant first class had the duty to study them to see what geography had changed (dirt being moved, indicating the VC were digging tunnels).

Any change indicating this would create a situation for the 11th Armored Cavalry. They would send two M113's and two rifle squads. They would work with two Dusters and one Quad 50 a sweep, then locate the tunnels, throwing in M56 grenades first and then smoke grenades to see where smoke would come out other tunnels. Then it would be necessary to use our machine's dozer blade to seal them up. Of course, in a couple of weeks they would be dug out again.

The briefing started with the CIA announcing that 600 of North Vietnam's hand-picked finest had been traveling for six weeks to come and isolate less than 200 of us including all Special Operations and Special Forces. The advanced party would be a squad of 12 sappers. They would be approaching the west wall.

The one Quad 50 was already covering the closed and locked gate, with the other on alert in the center of the compound.

My first night on greenline would be on 3 Alpha, next or close to the American gate. We all, all 30 of us, went into the bunker. The E-5 in charge of all of the greenline stepped in, locked a belt into the M2 .50 caliber machine gun, and pulled the charging handle once after checking everything on the machine gun.

After the sergeant left the bunker, Chief Silvers came out to the position.

He asked the Spec-5 in charge of the bunker, "Do you have a problem giving up your first shift and switching with my man here? If you do, I'll speak with your platoon leader and borrow you tomorrow instead. I'll let you feed the M60 on our way through Cu Chi and on into Cambodia tomorrow."

The Spec-5 remarked, "Chief, I would be more than glad to switch with him if you promise not to bother my platoon leader."

Chief Silvers stated, "It's a deal, trooper."

The chief then said to me, "These troopers here will be able to sleep in tomorrow morning until noon because of pulling greenline, however you need to get five hours of sleep doing two hours of watch and sleep four hours off here in the bunker; however, be ready to react to anything." He paused then said, "We all know what to expect tonight."

"Specialist McCormack, I will be here at 0700 hours. As soon as the debriefing is complete we will leave out. I will have a couple of the men set up the M60 in the Jeep and prepare for the trip. Be sure and have the poncho roll in the Jeep, web gear, AR-13 mounted in the weapon's rack you installed. I will bring a M79, and place a case of M56 grenades, and a case of grenades for the M79," the chief explained.

I pulled the first swing 1900 hours to 2100 hours. I then laid down to try to catch some z's. At about 1730 hours a new kid from the south screamed, "We got people in the minefield!" He rang the officer of the day, screaming into the field phone, "We got people in the minefield!"

The officer screamed, "Fire some grenades in the 79!"

The E-5 froze on the M2, stiff as a board. The newbie screamed, "I don't know how to fire this thing, I'm going for help."

I slapped the E-5 as hard as I could with the left hand in the back of the head and screamed loud as I could, "Fire that damned weapon!"

He was still frozen. With no time to get him out of the way, I fired a claymore, wrong one but another went off; the only thing I figured when first light came, no doubt I had set off two, one by accident. You see in our bunkers, there was a little wooden rack neatly arranged with a little piece of 3/4" board between the clackers (that is what you squeeze or push on, on our racks, to detonate the claymore mine). Safeties used by other troops were never used on Duster.

The clacker is nothing but a very small handheld generator. When it is squeezed, a larger gear spins a smaller gear, which rapidly spins the armature of the little generator fast. In my experience, it only takes one squeeze or stroke to fire it. Forget Hollywood, the back blast of the recoil can kill you, or harm you if you are behind the plate. It is armed with plastic explosive, and if my old memory still serves me, fire 244 one-quarter inch steel balls from the concave front plate.

This was my first time, of several more times, seeing this claymore wired to an engine stake with sand bags stacked behind the mine. The plates launched straight up for maybe 15 feet then fell back to earth with the two wire still attached.

We had fluorescent tape on the rear of the sand bags to know if the Viet Cong had turned them around and aimed at us, which became a common occurrence almost every night. Well, I ran wildly into them, as about half ran to my right and the other half to my left.

I had grabbed two belts of grenades for the M79 launcher. As I crested the berm, I fired a grenade wildly from the hip, missed by fifty yards but got their undivided attention. This scared the living purgatory out of me for about two seconds and then I turned wild crazy.

I ran to the edge of the minefield at the rear of the mines. Almost at that instant, the turbo charger whining on the Quad 50-gun truck backed in quickly but did not fire. I suppose because I already had that 12-person sapper team on the run.

From my first grenade I fired going over the berm, position 3 Alpha, about fifty to sixty yards or so south of my position, were firing sky flares for me as fast as that team could for illumination for my grenades to go out. I was firing a grenade about every three to four seconds as fast as I could. About grenade nine the officer on duty crawled up on the top of the berm, yelling, "What do you need?"

"Captain, get me some more grenades over the wall or the last one goes in your hip pocket," I screamed back.

In no more than ten seconds he was using 48 round cases to push the first, second, third and fourth cases down to me where I kneeled and hunkered down behind four sand bags stacked. That was some observer's position, which actually had never been used.

I kept pouring it on and about grenade fifty to seventy-five, I made a direct hit on one of the satchel charges strapped on one of the Viet Cong's back, which produced a chain reaction about 200 yards out "Boom! Boom! Boom!" At least three satchel charges detonated. Well, I fired 204 grenades, about the last 100 using the launcher like a mortar, butt plate resting on the ground. The recoil was so tremendous my shoulder could no longer take the force.

I fired dry, popped up and hauled my little butt back over the berm.

The sergeant major was carrying the kid who ran like a sack of potatoes by the waistband on his jungle trousers and the rear of his shirt collar. The sergeant major stood him up and said, "You run again I'll shoot you myself!"

I injected, "No you won't, sergeant major!"

"Why do you say that," he asked, looking straight at me.

I replied, "I'm faster than you. I'll kill this chicken shit piece of shit myself!"

Then the lieutenant colonel stepped in, "Trooper, never run into that minefield again, fight from the top of the berm."

The OD said to the lieutenant colonel, "I'm putting him in for a Bronze Star for valor for what he did."

The lieutenant colonel replied, "He was just doing his job, I will not approve the award."

Well, the captain put me in for the Bronze Star for about the next four months, three times, and I was turned down three times.

After I came back over the berm, I asked the sergeant of the Quad, "Why didn't you fire that thing?"

He replied with a grin. "Wanted to see what you're made of newbie; you did really good."

All personnel returned to their posts and it was quiet until a little after midnight. The Spec-5 accidentally discharged another claymore; I came out of a shallow uneasy sleep, sprang to my feet and he shouted, "That was an accident!"

Then the field phone rang in about five seconds, the E-5 apologized to the captain, "Accident, sir."

Well, I dozed off and a short time later I heard click, click, click. That dummy was playing with the safety while holding a M79 straight up in the air!

"Will you put that thing down, leave it alone and quit playing with it," I demanded.

He did not listen, and in no more than ten minutes he discharged the weapon straight up. What goes up will come down!

He came crawling in on all fours from where he had been sitting just outside in front of the bunker. He dropped the M79 in my face and flopped right on top of me.

I yelled, "I told you not to play with that weapon!"

About that time the grenade detonated about six feet in front of the bunker. We were all about three feet below and did not get hit, but the blast and concussion was bad. Fragments buried into the wooden upper structure of the bunker, and he began to cry like a baby that he would not live to see his family again.

I grabbed his shirt with the right hand and with the open hand of the left slapped his face three times and screamed, "Dry up you chicken shit pantywaist; you are going to get us all killed if you don't get your shit straight!"

I then said, "This M79 is mine. You are now not allowed to touch it; if grenades need to be fired, you wake me."

I did not realize until Chief Silvers told me the next day that the lieutenant colonel and the captain saw and heard what I did. From that point on, when I was on guard in the bunkers, lieutenant colonel and the sergeant major ruled to let the Kentucky man have the M79. That suited me fine. I had fallen truly in love with that weapon.

At about 0600 hours, two M113's from the 11th Armored Cavalry from Xuan Loc pulled up. With the 5-ton Quad 50 still in place, they asked the sergeant on the Quad to head them in the right direction.

At that time, the Quad 50 and two Dusters joined them. They spread out in twenty-five yard intervals, moving not more than a slow walk. There were twenty infantry troops eating from c-ration cans. The rear of the M113's had opened, and what you might call the tailgate was down and tarps and such spread on the floor with five gallon buckets spread across the tailgate.

At about two hundred yards out, the twenty-walking infantry carrying entrenching tools (small shovels) began scooping up internal body part, picking up arms, legs, severed heads and placing such in buckets and laying what was left of torsos in the rear of the M113's.

With 7x35 field glasses and standing on the top of the bunker, I watched that. I thought, "My God, I did that." When they returned after following the blood trails of six of them into tunnels and sealing off tunnels with a dozer blade, one sergeant asked, "Who did this?" While looking at 204 empty casings from the M79.

I admitted, "Well, I guess I did."

"Where you from back in the World newbie?" Sergeant asked.

"Kentucky," I answered.

"Well, you hillbillies sure know how to fight," he said.

I answered, "Guess so."

He replied, "You have six definite kills, three probables, and three maybes."

Nothing more was said as the two M113's moved away back to Xuan Loc.

Now at 0700 hours, it was back to the underground war room.

As I walked into the war room underground, a captain took me aside and stated, "The lieutenant colonel that commands the Tiger Rangers was in from the bush last night. He watched you and everything that went on.

"After the debriefing, there will be a ceremony to honor you… decorate you. I will stand behind you and whisper what to do, however, you don't have to speak. Just do what I whisper to you; we cannot afford to mess up and insult them, sit here next to me," chief told me.

"I will put this maroon Tiger Ranger beret on you properly and hang this patch over your right blouse pocket. Sit down now for the debriefing," he ordered.

We were debriefed about the fact that when the Quad 50 backed into the slot in the berm and the sapper squad failed in their attempt. The large force did not wish to face "Whispering Death" (the nickname given to the Quad 50 by the Viet Cong).

The enemy force broke up into small groups disbursed and went on to attack secondary targets. Now at this time our lieutenant colonel announced and introduced the lieutenant colonel commanding the 30th ARVN Tiger Rangers.

At this time sitting on the back row with the captain he whispered, "Stand up at attention."

The captain quickly placed the maroon 30th Airborne beret with the tiger flash on my head. He quickly hung the shoulder patch with black elastic over my right-hand blouse pocket.

The lieutenant colonel of the 30th took the floor. My battalion commander, the sergeant major, the major and some captains all stood facing me at the position of attention.

The 30th Tiger commander began to speak using broken English, "If all military men fight like 'ill billy from 'tucky, we would be in home in six months.

"I now honor you Mackie as my Tiger Ranger of my 30th Rangers," the captain said with all in front of me including the major, sergeant major and a captain of the 30th with all the Americans of the 5th of the 2nd all saluted me simultaneously together.

The captain whispered, "Return the salute." He counted off a number of seconds then said, "Release your salute."

I did. Then all facing me released their salute. We were dismissed. At that time the captain looked at me and said, "It is not a good idea to wear the beret and patch above ground. A sniper at eight hundred to one thousand yards will do his best to take you out."

"We now will have four hard days," said Chief Silvers. "Go to the hooch and gather one towel, one washcloth, a bar of soap, a razor with one blade and a tooth brush. Roll everything in the towel. And then come back and put it in the rear of my Jeep out of the way. I plan to lay over in Cu Chi for the night. We will be close to Tay Ninh for an early start in the morning."

I returned with the towel rolled and tied off with a thin piece of cord. The chief said, "While we are still here on the base, time for equipment check. Lock a magazine in your AR-15 but do not chamber a round. Two reasons: safety, and if a round is in the chamber for extended periods-corrosion and oxidation on the brass can leave all or part of a spent casing in the chamber. This is a major problem. Lock the bolt to the rear. Place it in the rack."

Continuing his orders, "Now take the ammo bag, pull the shoulder strap through with the upper sling mount and tie one loop where the bag just touches the floorboard. If we have to bail out fast, you will have your rifle and the ammo with just one swift movement." He handed me a muzzle protector and showed me how to place it over the flash suppressor to keep monsoon rain out of the bore. He explained that in an emergency I should just fire through the plastic. He said, "It won't hurt anything. I have plenty of them. When you need more, just ask."

The chief had mounted his M60 on a pole in the rear floorboard. We had one case of M56 grenades, and a case of high explosive grenades for the M79. He mounted a mini can of 1800 rounds for the M60. I had my poncho roll against the rear wall of the Jeep. We had earlier put three and a half gallons of gasoline in the five gallon can mounted on the left rear of the Jeep.

We pulled outside the gate and took off our flack vests and web gear. The chief placed his gear in the floorboard and advised me to place mine just behind my driver's seat.

He advised me to put a grenade in the M79 but leave it broke open about fifty percent for safety. I pulled it up against the mini-can welded between the seats that doubled what would be in the civilian world "a console."

We had no patches, nametag or US Army on our blouses. He turned his collar under and reversed his boonie hat to hide the officer's rank.

Chief then said, "When we clear this gate its free fire. Two rules: rule 1, there are no rules; rule 2, don't forget rule 1. We've got to look like the dumbest suckers in this man's army. It could make the difference of life and death.

"The Viet Cong are highly intelligent; never take anything for granted, that will most certainly get you or both of us killed." The chief added, "I don't want to have to write another letter to a loved one back home. I've had to write too many of them already. You see this is starting my fifth combat tour; I have had two tours of combat in Korea."

The chief paused a few moments and my thoughts went back to the previous night; I had crossed the threshold of taking human lives. Such a sad eerie feeling, something I never wanted to ever do, but it's done. One can never go back. I have learned over the years, that a man carries a feeling of guilt, no doubt for all of his days.

The chief began to speak and lecture again, "You must learn to scan constantly with your eyes sweeping side to side, watch the ditch line, watch the vegetation and trees, you are looking for any sign of a sniper, anything abnormal, a child in the edge that will detonate a mine. Be aware of any lighter or discolored dirt on earth. That subsoil is mixed with topsoil, indicating a mine is buried in the road. In that event take the shoulder, preferably to the far left if possible. If not, take far right. It will save your life.

"You must develop a sixth sense to detect danger in essence to literally almost smell it. I have that ability that has kept me alive so far. It is a God given sixth sense."

About this time, we pulled into the junction just outside of Binh Hoi. We were moving slow when two cowboys on a 50CC Honda pulled up behind us and in an instant, took the fuel can off the back of the Jeep.

The chief looked over at me and said, "That's my fault for not having you have the welders put a 3-foot chain on the can and locking it down. When we get back to Duster, I will issue a new can and the welders will weld the chain to the can. Next time you will have a laugh with those two."

In a couple of minutes, we pulled into the rear of the convoy. The major in charge of the convoy put us between the fuel tanker and the wrecker. The chief went off and there was a shouting contest.

The major finally said, "You are where I put you. This is my convoy!"

The chief yelled back and explained, "The fuel tanker and wrecker are primary targets for the Viet Cong. If that tanker takes a heat rocket from the Soviet RPG they will find nothing of our bodies. You understand, that it is a round that will collapse when it hits, burn a hole through the metal and eject molten metal ball bearings at several thousand degrees, and detonate that tanker and nothing will be left or remaining."

The chief then shouted at me. "We will play the major's stupid game, but when I tell you to move, do it with utmost speed."

We had gone only a few miles and the convoy took fire. The chief wide eyed, with the back of his left hand smacked the center of my chest, "Pass on the right shoulder fast, get us out of this mess."

I banged second gear, wound the little engine tight, pulled high gear with a speed shift, with full throttle to about 60 miles per hour.

The chief wide-eyed to the windshield with the right and the left hands holding onto the mini can console asked, "What did you do to this little machine to make it run like that?"

"Well chief, I just did a high-performance Corvette tune up."

"Good, I told you to do whatever you wanted, and you did good little Mack."

The chief had me stay 1/8 to 1/4 mile ahead of the convoy, pacing the convoy out front.

The chief said, "I hate convoys, they just draw fire; Charlie won't bother with two dumb looking privates when he has a whole convoy to shoot up."

As we were between Cu Chi and Phu Loi, we came closer to gunfire. The chief said, "This is too close. AK-47's and M60 for sure, slow down." Again, he slapped me across the chest, "Slow down; a rifle squad, at the edge of a rubber tree forest. We slowed down to a few miles an hour over a hill and a curve and there was a rifle squad of about ten men walking on line five meters or so apart. Silvers screamed, "Korean White House Marines!"

Two Viet Cong were on their knees, hands out in front clasped together, begging for mercy. Two of the White House Marines shot them in half with the M60 machine guns.

Chief Silvers said, "The White House Marines take no prisoners!" There were two Viet Cong running parallel with us beside the road. The chief grabbed the mic to the radio and screamed into it. About 30 seconds later, a jet dropped down to little more than true top level and dropped napalm no more than 30 to 50 feet to our right side. The chief ahead on the M60 gave both of the Viet Cong a quick shot each with the M60.

"It is the humane thing to do!" Chief said.

"Drive, drive, drive, get us out of here! That napalm will rip the oxygen from our lungs!" the chief screamed.

I poured on full throttle banging gears, with headlights on as it was dark getting through the thick rubber trees.

After clearing the trees, the chief was back on the radio letting the pilot and the Republic of Korea Marine Corps. The ROC's know the napalm and his M60 took care of the last two Viet Cong.

We arrived via the south gate of Cu Chi, pulling up to the gate. A private first class jumped all over us for not wearing helmets and flack vests.

The chief stepped out of the Jeep, turned his bush hat over revealing the rank of chief warrant officer 3, pulled his collar out revealing also his rank.

"Private first class come to the position of attention in my presence as the officer I am. Respect the rank!" Chief Silvers commanded, very loudly.

"Private first class," Silvers said as he took out his large lock blade knife. "Do you want this button?" indicating a blouse pocket button.

"Sir, no sir!" the PFC replied.

The chief cut the button off and tossed it aside in the grass.

"Private first class, do you want this button?" the chief screamed louder.

"Yes sir, I want that button," the PFC shouted back.

The chief cut that button off and stuck it into the pocket of his shirt.

The chief pulled a small note pad and pin from the blouse pocket. He opened the pad and prepared to write.

"Private first class, I want the name of your sergeant major."

The PFC answered.

The chief jotted down the sergeant major's name.

"Private first class, I want your battalion commanders name."

The PFC answered.

The chief wrote down the commander's name.

"Private first class, look at this black Jeep. It belongs to me. This trooper works for me."

"When this Jeep pulls up to this gate, log it in, keep your mouth shut, and wave it on. I will not tolerate any more of this petty crap. Understand PFC?"

"Sir, yes sir!" the PFC answered.

"You mess with my vehicle or my trooper ever again, you will no longer be a private first class, I'll see to that! Understand private first class?"

"Sir, yes sir!"

"You are dismissed," the chief ordered.

In the days to come that PFC and I became friends, thanks to the chief.

We drove deeper into the heart of the base camp, as I was shown by Chief Silvers. There was an underground bunker to drive down into. It was designed for vehicles, with overhead protection in the center.

The chief commanded, "You stay right here underground with the Jeep. Do not wander out of this bunkered position, stay with the vehicle until I return, understand?"

"Sir, yes sir," I replied.

I stepped out of the Jeep and climbed up on the front bumper so eye level for me was just above the ground level.

Chief Silvers was returning to the Jeep when many mortars and rockets were coming in. Silvers froze, trying to run but his legs froze. His arms were being thrown around. He was turning his head in all directions but his body was frozen or locked up from the waist down. The attack lasted several minutes. People were hit all around him.

The most uncommon valor I saw or witnessed my whole tour was on that day at that instant.

Several nurses, ladies in nurse whites, with the aid of two corpsmen were checking the men that were down. They were checking their pulses and heartbeats, placing them on litters and racing in a ducked crotched manner, rushing them to the emergency room.

They checked on one man about thirty years or so and left him. The sunlight was gleaming from the wedding band on his hand.

The attack ceased and the chief rushed to get back to me and the Jeep.

I was still standing on the bumper when I asked, "What about him, chief?"

"He is gone; I know him from Knox. He has a wife and two children waiting in Radcliff."

"Well, he will be processed through graves registration and be back home with his wife and two children soon enough. In a controlled voice chief said, "The widow will draw the survivor's compensation from veteran's administration for as long as she lives and as long as she does not marry again. The children will be well cared for."

The chief was whispering like we were in class, "We will sit here for a few minutes to be sure it is all clear before we leave. There is a small convoy going into Cambodia, we will follow them to a staging area." After a moment of thought, "I need to talk to a couple of the Duster crews to see what is needed so I can respond to their needs."

It wasn't long before we fell in behind the small number of supply trucks. The chief instructed, "Stay back one hundred yards or so, if they draw fire we can make a quick U-turn and retreat from an ambush."

The rest of the trip was uneventful all the way to the staging area; however, when we crossed a pontoon bridge over a small muddy river the chief remarked, "We are now in Cambodia."

About maybe 12 to 15 miles down the road, we pulled into the staging area. Four M113s held the four corners, a man manning the M2 .50 caliber machine guns on each M113 track machine. The four rifle squads were lying prone; each man was on line with the M113s in each corner. The chief spoke with the two sergeants, track commanders for the two Dusters refueling and re-arming (this term is used for the task of filling the ammo boxes with fresh 40 mm striper clips, and filling two very large ammunition wells behind the driver's seat and behind the assistant driver seat down in the track).

I spoke out, "Anyone here from Kentucky?"

"Yes, preacher up in that gun tub on that track is from somewhere in Kentucky," one of the troops answered.

I climbed up to the gun tub and asked, "Where are you from in Kentucky?" He looked up from reading his Bible, and said, "Not so far away from Paducah."

He dropped his head and began reading his Bible, so I did not bother him anymore. I climbed down off the track and went past the sergeant who said, "He just doesn't have any fun, he doesn't drink, smoke or fool with the ladies; all he does when he has what little free time we have is read the Good Book."

The chief spent a few more minutes with the sergeants and declared, "Time to head back to Cu Chi for the night."

The trip back to Cu Chi was uneventful. We could still hear constant gunfire, but all at a distance.

As we got closer to the main gate of Cu Chi, the chief spotted the ten-ton wrecker with the forty-foot lowboy. There was a M113 with a rifle squad with the driver of our "88 on wheels."

The chief threw his fist against my chest, "Pull in here next to my wrecker, look at this shit, they are partying and drinking beer. He just as well give his heart to God! His little ass belongs to me here and now!"

I stopped the Jeep next to the wrecker. The SP2 dropped his can of beer, jumped up coming to attention. He attempted to raise his hand to salute the chief, "Freeze Boy!"

The specialist froze.

"Boy, you salute me out here in the open and I'll slap you so hard your face will trade places with your Pillsbury ass! Got that, boy?"

The specialist replied, "Sir, yes sir!"

The chief screamed, "You get your little ass in gear, you get my retro grade Duster on my lowboy and be back on Duster Compound by dark or God as my witness I will put your young ass in Leavenworth making little rocks from big rocks with a sledge by hand." The chief paused and shook his head and continued. "If you ever pull this kind of shit again your head won't even have time to spin! Understand me boy!"

"Sir, yes sir!"

"Now reach up, scratch your nose, you dumb ass, then get your young ass in gear, dismissed! Go boy go," the chief commanded.

The chief returned to the Jeep, climbed in and just sat a couple of minutes staring through the windshield saying nothing. Then he broke the silence saying, "You know, if word got out to a star, I could be in real trouble and screw up what I have left of my career."

I commented, "Chief sir, I did not hear a damned thing, must have been just the wind."

The chief finally smiled and commented, "Thank you, little Mack."

I replied with a grin, "Chief sir, I rather enjoyed seeing that, that was better than watching a skin flick."

The chief's mouth kind of fell open, such a solemn look on his face for about 60 seconds then he did something he would seldom do in the future. He burst out laughing. After a good laugh, he said, "You are going to make it just fine, little man, just fine."

When we arrived, and checked in to the line battery headquarters, the chief learned there would be a major ground offensive coming that night.

The chief said, "We are not staying here tonight. There is major ground offensive coming, I had enough of that in '68. We are headed back to Duster Compound for the night. We can resume early tomorrow." The chief looked at me with shitty grin and said, "We have three more days together before I turn you loose on your own."

All the way back to Duster Compound the chief was lecturing and teaching me the dos and do nots of the job. Most of the time I would be traveling alone to reach firebases where the machines were spread about the southern third of the country working with the infantry; understand, the 5/2 did not fight as a unit but from one to five machines with infantry units fighting with them with E-5 sergeant in command of the crew and machine.

The next three days we traveled up to Tay Ninh, the northern point of the Iron Triangle, and several firebases visited Phu Loi and Xuan Loc. What was most eventful were the seven stops and being introduced to seven covert spies, local civilians. For each one, Chief Silvers would slip a roll of piasters currency while introducing me to each. In the end, three of the spies were not dependable. After visiting those spies and being given a sheet of paper folded up very small, placing such in my boot, I was told to not return to them. Four of the group were very dependable and trustworthy. However, I always hung my AR-15 slung from the front when I would exit the Jeep and walk a few feet to greet them, slip them a roll of currency and covertly receive a sheet of paper folded about two inches' square.

One of the dependable spies was an older Frenchman. He had been a major with the French Foreign Legion. He owned 3,500 acres and employed many families, each farming 20 acres each. He had six ex-Legion NCOs in his employ. They wore camouflage battle dress and carried sub-machine guns. They stayed close to his house during the daylight hours. He had a very nice old French staff car and lived in a three-story mansion. The ground floor had bars on all windows and very heavy front and rear doors. On the second floor sitting in a doorway out to a balcony stood an older French World War II machine gun on wheels. On the third floor, out on the balcony stood an old cannon about 75 millimeters.

He explained that the six men were his security force to get him to the airport and leave the country if the Viet Cong came after him. He, his young wife, and little son would be evacuated if such happened.

The men worked during the hours of darkness; three would stay with him and his family and the other three would recon the villages on his plantation to observe which of his tenant farmers were Viet

Cong. They would return to his house, give him such information. He would put such to pen and paper, slip it to me folded up about two inches' square slip it to me in a hand shake and I would slip him a rather large roll of currency. He informed me he paid taxes to both the Saigon government and to Hanoi and to leave him alone and not bother him.

The next spy was a Buddhist monk. The first time I visited him to exchange the money for the folded paper he spoke to me in English, "I do not like you Americans but hate the Viet Cong. I only do this to buy food for these people here in the leper colony to feed the lepers."

Most of the men in the leper colony had hands removed and many had nothing below their knees. They were helpless. The monks would feed them and even carry them and sit them on the open toilets.

Just to handle that folded piece of paper after the monk had handled it, and handled the leper, sent fear through me that I too may catch the disease.

The third spy was a sweet young lady at a Y intersection that washed clothes for whoever stopped. People left jungle fatigues and Tide laundry soap from one of the PXs, and some paid her a small fee and returned later to pick them up — men in tanks, Dusters, armored vehicles and such.

I would stop, leave a set of ratty jungle fatigues furnished by the chief with her money in a pocket. I would give her a box of Tide to look good, as the Viet Cong would most likely be watching. She would slip me the folded paper in a handshake and I would slip it in my boot the same as the others in the same manner.

The fourth spy was deep south on a dirt road at the edge of a smaller straw hut village. There was a road side little lean-to of buildings that sold tiki torches, candles and whatever he could get his hands on. He would stick the paper folded up on the bottom of a large candle. I would slip him the roll of currency, place the candle in the mini can between the seats, being careful carrying it bottom down in the palm of my right hand placing it in the mini can and locking the pad lock for security. I was always very uneasy and fearful when I visited any of the four of them.

Returning to the hooch, I received a strange letter from the 603rd Transportation there at Polk; strange, I didn't know this man all that well. I opened the letter and it read like this:

Mack,

Your girlfriend you bought the ring for and are engaged to, well she is running with a first lieutenant that just came back from Vietnam. He was in a grunt unit. He is driving a red Corvette Stingray coupe. She is trading sex for dope. They park under the weeping willow tree where you spent time with her. Sorry, but you needed to know.

Your friend,

Well that it hit me hard for a moment. Then I thought, well it's over for me and her. Oh well, I have a job to do and a war to fight. I figured I didn't know if I would get out of there alive, but I knew that I would do everything in my power to see others get out.

Just as my heart took a hit, I would soon adopt an orphan in a hamlet in Honi. Each enlisted man gave $6.00 a month to feed an assigned orphan. My little one was a four-year-old, likely from the northern part of the country. Her mother and father no doubt had been the enemy.

My first Sunday evening break, the last Sunday of June, I met my new temporary baby. This was the first break I had since entering the country. The priest called her over as all of the children were playing together. She ran and jumped in my arms. I asked the priest why she had no shoes or sandals. He replied, "There is no money."

I abruptly asked, "How much money?"

He replied, "So expensive, 18 U.S. dollars."

I reached into my pocket and brought out my last $20 bill. I put it in his hand then stated very strongly. "Buy her a pair, the remaining $2 buy chocolate candy. When I return she better have shoes."

About three Sundays later when I was free to see her again, she ran and jumped in my arms hugging me around the neck so tight, choking me. The priest told her not to choke me. I learned it was her first pair of shoes. The priest translated for her, "Thank you so much my papa, I love you so much."

I stated, "I love you too baby."

I learned all of the children were given a little piece of candy. I also learned the children never could have meat. I made a deal with the priest: I would give an extra $2 each month so my little one could have a little fish or chicken about twice a week. After she experienced a couple of ounces of pork meat, a little chicken and fish, she commented to me on a later Sunday, "Papa, the meat is so tasty."

I learned that day the international Red Cross furnished the children's clothing. I could tell by holding her that she had no panties; none of the children had underwear. I walked to my Jeep, took out a piece of string measured her little waist, and tied a knot to the correct waist size.

Back on the base I used a tape measure. I wrote mom that night, as mom had a power of attorney on my bank account to go and buy several pairs of little girl's panties in Elizabethtown and send them ASAP.

I also learned she had only one little dress. It took some days to get her a new one. She was so happy, it was a first for her and the first would have been for any of the children.

All of the officers and NCO's gave a little money each month for the few crib babies, and for French books, tablets, pencils and school supplies for the little ones as French nuns cared and cooked for the children.

CHAPTER FIVE

O.K. Corral

Waking at first light, it was new day! My surprise training ended five days ago; today is a new adventure. I will go it alone today. Everyone on Duster was becoming accustomed to the machines that rolled in every morning and that I was decorated in the underground war room.

At about 0645 hours, a convoy of artillery field pieces came roaring through our gate. A self-propelled 105mm was running point. Following them were 155mm and 175mm field pieces. Behind them was an old 3/4 ton with a mounted 75 recoilless rifle. There were two 81mm mortar teams with their trucks. They had spread out the entire southern berm on this little compound. When I left the hooch, I stopped higher on the hill and noted the two Quad 50's sitting, and one Tops Popcorn Machine still on the gate. There were others in the center of the compound still in ready reserve mode.

As I scanned, I noted as usual one Duster respectful in the southeast corner and another in the southwest corner. A third Duster centered among the freshly brought in field pieces. A fourth Duster on the east berm was on ready reserve as well. It rotated regularly with the three that were firing semi-auto, one-gun tube only at a time. I would learn this was being done to keep from overheating the gun tubes and burning out the bores, which would harm accuracy of the gun tubes.

Each team fired their own gun tubes, rotating the machine in a 180-degree sweep, the rounds going out on the high plane every three to five seconds. It was at first difficult to get a little sleep.

I took a couple of minutes to reflect back to the 11 months there at Fort Polk, living in those old outdated WWII buildings with no air conditioning.

I thought of the riot control training and extensive bayonet training in that hot humid climate. The hand to hand and karate up to brown belt ten. How to take down an unruly rioter, put him down, beat him into submission, roll him over, and put restraints on him with hands behind him.

I thought of the disarmament, pistol revolver, pistol automatic, long gun, disarming the knife or club. To then put him down and restrain him. It was good training for the situation I'm in now.

I dearly love this new job of becoming the driver and courier. To get off this little base, being on my own alone. To literally be my own boss, just get in wherever necessary and get out again. To have been able to modify that engine my way with Corvette specifications, what else could I even ask for? Nothing.

I was working for one man on his fifth combat tour who had won multiples of every award possible with the exceptions of the Congressional Medal of Honor, a man so intelligent and brilliant that he lacked only three courses having his doctorate at the University. Having to participate in this war thing, well I could not ask for anything better for me.

Well the chief just entered his office. Time for this little man to report to the finest officer in this man's Army.

I reported in front of Chief Silvers in his office. I saluted, and he quickly returned a one second salute sitting behind his desk. He glanced to Captain Ben and commented, "Captain, I need to speak to the specialist alone, please close the door behind you and cover the door as no one disturbs us for a couple of minutes."

The chief sat for about a minute silent then spoke, "At ease little Mack, we are alone now as we were these last four days. When others are present we must have military protocol. However, when it's just the two of us, it will be like now, just you and me, it will be just chief.

"Now day one in this office I gave you instructions not to discuss anything with anyone else, the men in the platoon, or anyone else about any of the training you were doing with the Rangers or the discussion of anything of the people you will be visiting." Chief looked me in the eyes, "You haven't disclosed any information to anyone, have you?"

"No sir," I replied.

"Good, if you had or do, it could very easily cost you your life. Understand, there could be young men here that may sell you out for a tumble with one of the females in the village.

"Now come Saturday, I will accompany you to the village to introduce you to one farmer that can be trusted to buy vegetables for my platoon. I will each Saturday give you money to buy ice at the village and vegetables from only the one farmer. I will accompany you to Long Binh to buy 13 T-Bone steaks and 13 baking potatoes."

"Russ will be relieved at 1100 hours on each Sunday to grill steaks and potatoes, a good meal for my platoon once each week. In the event you have tower duty on Saturday 0900 hours until 1900 hours on Sunday, someone from the platoon will bring your meal and you can raise it in the bucket with a rope. I hate to have to do that to you, however I need you on week days to make my runs.

"Remember this, no one else touches that Jeep with a wrench except you," chief ordered.

"Shady work on that machine could leave you stranded in the middle of nowhere, and with the training you received you know where that will leave you."

"Yes sir," I replied.

"I will never from this point on order you anywhere," chief said. "If you find you cannot get through turn around and get back here, understand I need you back, understand?"

"Yes sir," I replied.

"As we speak, my strip crews are pulling a breechblock for a gun tube, fuel pump and Magneto for the Dusters on a firebase on the way to Cu Chi."

"Remember that sharp hard sweeping left curve with the sand trap on the right between Cu Chi and Phu Loi?"

"Yes sir. You pointed that out as being somewhat dangerous," I replied.

"Okay, that sand road that turns right at a 90-degree angle has a large firebase 28 miles out. A Quad 50-gun truck will meet you at the intersection at ten hundred hours this morning. There are four Dusters there that need these components."

"One-gun tube is down on one of them, and two machines are immobile. Together you and I will keep them firing and moving, which will save many lives."

"That is our main concern, to save lives by those machines killing as many communists as possible to keep them from killing our people." The chief commented again, "As I told you day one, I know your father well from transportation motor pool at Knox, which is my home base. My wife and children are awaiting my return there. Your father needs his only son back after this mess is over.

"Did you know he still carries a photo of you in his wallet when you were a small child, your sister too?" Chief asked.

"I guess so chief," I answered.

"Now check the oil in the machine. Strap that knife on; never be without it when you leave this base. Check your gear, you know the drill. Army Ranger, I need you back, remember your training. Go now!" Chief commanded.

The Jeep was prepared, loaded up and rolled out of the compound. What an adventure.

The little machine stopped at the gate, I reached down, locked the bolt to the rear on the mounted AR-15, and popped in a twenty-round magazine. I reached in the rear, grabbed the 79, popped in one grenade round. I left the breech half open. I situated the bag with my six M56 grenades, with the shoulder strap, well just in case of a quick bail out. I ran a forefinger under the strap of the combat knife and pulled for a little slack as it was a little tight. The gate guard raised the pole. I was off and running. On past the old French fort, up on Highway 1, and on over onto Highway 13 named "Thunder Road."

Thoughts went back home before the war. The old movie made with Robert Mitchum portraying Luke Dulun of "Bloody Harlan" County, Kentucky. He had moonshine in his 1950 Ford two-door. The big Cadillac ambulance engine souped up with three 2-barrel carburetors. A racing camshaft, and exhaust pipes rumbling as the old coupe rumbled out of the holler, with 250-gallons of moonshine in the trunk. Truck springs in the rear to keep the tanker sitting level. Federal agents trying to run him down.

The stick shifted into overdrive, maybe turning out 140 miles per hour on the flats. The man that could roll that Ford over three times come up with the rubber on asphalt, do a fish tail 180 degrees and leave the federal boys like they were standing still on Thunder Road. Sometimes into Asheville, sometimes into Memphis town, the law they tried to catch him but the "Devil" got him first.

Banging up my "Thunder Road," thoughts went back to the steel grey 1957 Chevy two-door, the kids on the street, called the "Grey Ghost."

I thought of the six weeks with no job. An old black lady on the hill, I called Maw, gave me a part time job hauling cases of government taxed half pints from the junction in a wet county to the neighboring dry county with no liquor stores. You see, the old lady was surviving on $87 a month on her deceased husband, so on the side she bootlegged half pints of Kentucky bourbon. It was just a violation of a county ordinance, no serious crime; well the locals needed their little drink, too.

Well, I had made a run a week for Maw and things had been easy and quiet until the law tried to run me down. Five cases of bourbon strapped down. The rear seat pulled: three cases strapped to the floor, there were two cases in the trunk strapped down with added on seat belts for high speeding to keep from breaking the glass bottles. Well this time trouble, when I topped a little hill on what was called Bloody Ridge there he sat, that big Ford 4-door with lights across the roof. My Chevy grumbled by at low speed. That big Ford body twisted and hit the ignition and fired up that big 390 injector engine. The flashing roof lights came on as the '57 was cresting another hill.

Grabbing the Hurst floor shift, a double clutch to low gear on the 3-speed transmission, revved up to 7,200 revolutions in low. Power shifted to second after rounding a dead ninety-degree turn to close to 7,000 rpm's in second gear. A double clutch back in low after hard breaking, another ninety-degree turn onto a farm lane by a little church house. Winding to low gear again to 7,000 rpm's.

I shot into a cornfield of dead stalks and shut down the engine. Looking in the rear floor board, the Stevens double barrel 12 gauge had been sliding around, and a couple of the Peters 00 buck shot shells had started to crawl up out of the elastic sleeve on the gunstock. Well now that old shotgun was just insurance so thieves would not steal Maw's half-pints. It had been tried on other haulers. Now after sitting quietly for about 45 minutes, that little 270 horsepower 283 V8 fired up and would take the alternate route already worked out.

It was back to the junction, through Lyon's Station, through Hardyville, and I came out south of the hometown going north on 31 W into town, then grumbled up Glendale Hill to Maw's little sawmill house that sat on top of the hill.

The old '57 Chevy backed up to Maw's front porch. I could smell the aroma of pinto beans and hamhock that Maw had cooking on her wood stove. I was hoping for a taste when Maw stepped out on the porch with a walking cane in hand.

"Boy, where you been? You should have been here hours ago, I done give up," Maw shook her head. "Figured they done got you like they got my nephew. They still got him in that jail."

"Well, I had to rely on the alternate route through Hardyville and come in from the south," I answered.

"That crazy fuzz that got my nephew, 'bout got you I'll bet," Maw said.

"It was close, Maw," I said as I helped her carry in the liquor. She put it in a hidden closet she added after her husband passed away.

"You can't haul for me anymore, not anytime soon, anyway. Here is your twenty-five dollars," she said handing me the money.

"You sneak back the way you came in. Every Johnny Law is gonna to be watchin' for that Chevrolet. Take it somewhere and hide it. You will have to walk or hitch rides for a while," Maw said shaking her head. "They catch you, you will be rooming with my nephew."

"Well Maw, I'm going to the Army real soon. Guess if they want me they will have to come for me. You think maybe if I land in Vietnam they will come there for me?" I asked with a grin.

"Not unless they are plum crazy," Maw laughed, showing the absence of teeth.

Leaving Maw, I headed south, taking the alternate back route. I hid the Grey Ghost in a friend's father's barn for a couple of days and slipped it back home. The patrol cars sat across the street 24/7 for six weeks waiting for it to move onto the street. What they didn't know was I was off doing basic training at Ft. Knox.

I knew if I had been caught the system would have literally put me under the county jail. Now back to a reality, and back to the war right then and there.

Coming up to the curve, I was early — no Quad 50 here. "Oh Shit!" I thought.

The Jeep made two wide circles about 50 feet across. "Oh well, I can't keep circling or Charlie will kill me for sure." I started in on the sand road, going faster up to 50 miles an hour. The Jeep was starting to get a little squirrely. I dropped to 48 mph and she stabilized. This is the first time on sand for this kid.

I thought that it looked a lot like Kansas that I had seen on TV back in the World...the bushes and all, even like east Texas west of Dry Creek, Louisiana. This land is straight with hardly no curves to amount to anything.

After a time, I could identify a few shots from what no doubt were 40mm Duster rounds. I could see in the distance the coils of wire across the path I would enter. I picked up the radio microphone and called the firebase by name. "This is Duster Chief, I say again Duster Chief, pull the wire I am coming in."

The reply was, "Got your lima-lima Duster Chief, will do, over and out."

I spotted two Dusters setting about 50 feet apart, both with the armor plate removed from the rear decks. I saw the Quad 50 getting ready to leave to meet me. The Quad 50 pulled up next to the Jeep, "Hell! Duster Chief we just saddled up to come for you, you are damn stupid or crazy to come through the territory you just came through! The damn Viet Cong are everywhere." The driver of the Quad 50 was a big stocky 6 foot or so Duster crewmember of the gunnery crewmembers. He climbed down from the Duster on the left, came walking over and said, "Newbie, you are crazier than Hell to come in here alone, damn wonder your ass is still alive coming in here like that!"

"We killed the hell out of them last night. We lost some of our grunts last night as well."

"You brought my fuel pump set and my mag distributor I see; I'll get started on putting them on my machine," he said.

At that time, a little buck sergeant came into view. He was about 5' 5" and about 135 pounds. "Where are you from back in the World," he asked.

"Kentucky, just south of Knox," I answered.

"Come over here and talk with me. I want to talk to someone that still has their head screwed on straight. These guys have been here too long; their minds are wasted. We're close neighbors. I'm from Tennessee," he offered. "How're things going back home?"

"Same old, I guess," I replied. "I'd love to show you the '55 Chevy dad and I built when I turned 16 in high school."

"I went to the local tech school and took auto mechanics. My dad is a very good auto mechanic," the sergeant added, showing me a photo.

"This machine is beautiful," I said after examining it. "Hey man, those are C series Carter AFB twins!" I said with excitement.

"Yes, it's a 265 bored, .060-30-30cam, 375 horsepower." He added, "Warner T-10, close ratio 4-speed Hurst competition, plus shifter and 4:11 gears with line locker posi-traction."

I asked, "This '55 are mass wheels' Keystones?"

"Yes," he replied. "We run D modified production at the local drag strip. I always drive pulling 8,000rpm power shifts and always win my class. There have been times when I took top eliminator," he explained.

Showing me another photo, "This is the lady I'm engaged to. She just graduated high school, and I could not be there for her graduation." He stared at the sweet smiling face in the picture. "Stuck in this hellhole and I still have some months to go." The sergeant continued to stare at his lady.

Putting the photo back in the protection of his pocket he asked, "Hey, have you eaten today?"

I replied, "Na, I missed the shit meal at Duster Compound. I can go down to Phu Loi for chow at the Special Forces mess. The chief worked that out for me a couple of days ago, and said it's the best chow here."

"Nonsense, I'll get you a mess kit and feed you here from the hot meal brought in," the sergeant insisted.

"Nah man, I don't want to short you guys," I replied.

"Fella, they haul a lot of it out of here and sometimes after its been here 24 hours just give it to some of the village kids to keep from wasting it," he said.

I took the mess kit and it contained hot roast beef and mashed potatoes. When I finished, one of the Duster crewmembers cleaned it for me in the galvanized barrels with heaters under the can.

After eating, a Duster crewmember went to work filing the breaker points in the mag distributor. The crewmembers gathered around me.

"You have to have a good nickname. The sergeant is Wyatt. I'm Virgil. He is Morgan and the one sitting in the gun tub is Doc, from the O.K. Corral, you know.

"Hey, we will give you the nickname "That Hillbilly Kid," one of the crewmembers said.

The SP4 from Arizona said, "No, hell no, we are going to name you "Billy the Kid."

"Why is that?" I asked.

"Billy the Kid was totally crazy." He paused, "You are also totally crazy for coming in here alone without the Quad 50 escort." The SP4 replied, "Billy the Kid had two guns, the lever rifle, and his pistol. You have the rifle and the 79," he paused smiling, "Two guns."

"Both of you, crazy both, had two guns," another crewmember added.

The crew was in total and absolute agreement. "Billy the Kid it is. We choose the nicknames newbie; it's our choice."

I laughed, "Well, it's a good choice, we both have killed bad guys."

"You done got a kill, have you," one crewmember asked.

"Hell yes, the ARVN colonel even decorated me with the tiger patch on my pocket and a maroon beret for blowing away a sapper team on the 28th. They were trying to get on Duster Compound."

"No shit, how many," one of the crewmembers asked.

"Six confirms, three probables, three maybes," I answered.

"Damn, you have some of these newer grunts beat, man." The sergeant from Tennessee said, "Us hillbilly's do two things well, making moonshine and killing."

We all laughed.

"Well fellas, if you got what you need I need to go," I said.

"Hell no! Not without the Quad escort," sergeant declared.

"Aww don't worry," I said.

The sergeant screamed, "Captain, come here! If this man tries to leave without the Quad arrest him, and put him in irons until the Quad can escort him."

The captain said, "Are you going to wait, or do I slap irons on you?"

"Well sir, I won't leave until you give me the all clear," I said.

"That's better, you need to listen to these old timers," the captain smiled.

"Yes sir, got the message," I replied.

The crew started asking questions about my life back in the World.

In about 15 minutes the Quad 50 fired up and pulled up beside the black iron horse and the sergeant said, "Hey Mr. Billy the Kid, we'll take you out now, Charlie won't bother you with my whispering death escorting you out. Let her roll kid, we got your back." The driver waved me on.

Out we went without incident. I went on through the rear gate. The PFC signed me out and waved me on. In the days to come, the PFC would see me coming in, pop up the bar, wave me on. I didn't have to stop. Within days the entire legend of "Billy the Kid the lunatic" was on every track crew's lips, the Quad crews, the searchlights, and any combat armsman. All the base camps were talking about the hillbilly "Billy the Kid." I earned the reputation that the grenade launcher was my outlook and death did not faze me. With each mission my reputation grew.

I walked into the line battery office and the clerk stated, "Chief Silvers has had some Duster components flown up here for you to deliver. The lieutenant colonel on Duster Compound and Chief Silvers have given orders that our captain trace these maps. Captain will give them to you to see if you know where you are going."

The clerk reminded me that none of us wanted to wind up in Leavenworth for something going wrong.

The line motor pool mechanics loaded breechblocks and other used components in the Jeep. The captain called me in, "Trooper, come into my office. I've traced maps for you. I need to be sure and get you in the right places."

After briefing me, the captain said, "My clerk will take good care of you from here on out. Anything you need, just ask; you need gas, my motor sergeant will take care of you. God go with you, son."

I was rolling south to the firebases that badly needed the components. I finished my day and headed back the 72 miles to Duster. I thought how I loved my job, just me and my guardian angel that God had sent down just for me.

About a week later, the firebase I pulled into had the same track. "Hey," I thought, "Here is my friend from Tennessee, I see the number on his track. It will be so good to see him and talk to him again!"

I hurried to the track. I saw the kid from Arizona, and quickly asked for my friend. He dropped his head and softly spoke. "He went home."

I spoke, "His time is not up yet, emergency leave, was it?"

"He's DEAD-DEAD-DEAD-DEAD!" he screamed.

"Where do they get you newbies, you don't know nothing, just nothing!" he cried out.

I was in essence hit hard, a shock like being hit with a mental sledgehammer. Just the first hit of what would be so many.

Later on, during the tour I would kill three more VC there on Duster during the hours of darkness on the greenline. I began to take to heart and think about one of the things that Silvers had talked with me about. He had asked, "Do you believe in God, how do you feel about that subject?"

I answered, "Oh yes, chief."

The chief came back, "You know I grew up in a Holiness type church, and I believe that is what has kept me alive this long." The chief was quiet for a short period of time, and then said, "It's time to teach you a soldiers' prayer. It should go something like this, and one should speak to God about every 12 hours.

> *"God, thank you for letting me live these last 12 hours. God, if I must die, make it painless and swift. God, I don't want to lose major parts of my body. God, if you allow me to live, I do not want to lose the ability to have children of my own someday. God, every soldier's job is to kill the non-believing enemy, help me do this and guide me through the next twelve hours. Amen."*

The chief rode quietly for a time then asked, "Did you memorize the twenty-third Psalm as a child?"

I answered, "Well, kind of, chief."

The chief replied, "I suggest that for the sake of your soul you brush up on it, and when the shit hits the fan recite it as you fire your weapon."

"Yes chief, I will do exactly that," I agreed.

The chief went on that day to speak of his four previous combat tours. The first Korean War combat tour wasn't quite so bad in the early days of that forgotten war. The chief said, "If you stayed sober, left drugs alone, kept your grades up you had a good chance of making it."

"Chief, what do you mean by keeping your grades up?" I asked.

"Well, stay and be awake, aware, alert, agile, and you have a good chance to stay alive," the chief replied.

"Yes, I now see what you mean sir," I replied.

Early on, the chief had me take him to a neighboring camp, Bear Camp, to check with a gunnery crew with a new experimental weapon that would replace the Duster if the war went on.

We pulled into Bear Camp and drove to the east wall. There was a machine, a deuce and a half with a multi barrel chain gun mounted on the truck bed. The chief walked over and began speaking with the gunnery crew commander.

It seems that crew had trained at Fort Sill, Oklahoma, not only the Duster and Quads, but had been holdovers. These are people held over after the class had graduated and went on to Vietnam to one of the battalions but they had started a new training on the Vulcan, even experimenting and learning new tactics, and developing new procedures that would be common practice in the event that the war went on and the Vulcan replaced the twin 40mm twin gun tubes.

The chief would recommend to the Department of Defense that the same chassis of the M42A1 be kept; however, that the gasoline engine be replaced with an upgrade diesel, as the gasoline engine was dangerous and a fire trap hazard. Then there is the fact that the two gasoline tanks located on each rear corner of the tracks were prone to fire, especially from a shaped heat from the Soviet 40 mm RPG. In such an event, I learned from the gunnery crews that one had to bail straight over the front and run like purgatory when such set off a fuel cell.

I walked over and began a conversation with a SP-4 of the Vulcan crew. The SP-4 began to explain that all of the radar and components had to be crap canned and the old-world war sights from the twin 40's, that were mounted between the 40 mm gun tubes was mounted to the right side of the Vulcan gun barrels and visual fire system was used.

I asked about the rounds in the continuous belts. The SP-4 went on to explain, "We have one armored piercing, then one tracer, then one high explosive round. The Vulcan has been modified with a hand operated variable rheostat to control the fire rate from 1200 rounds a minute up to 6000 rounds a minute.

"We keep it normally sitting low at 1200 rounds a minute. We can feed that rate, but much higher the weapon will go dry because we cannot keep up feeding that hungry thing."

I looked to the east berm wall. There were dog kennels built into the wall. The kennels were about six by eight feet in size. They were about four feet tall. The top and outside to the open terrain had just bars. Each kennel had sort of a dog house to have protection for the dog during daylight hours while this period of time the dogs would return from the high plane from hunting the Viet Cong during the hours of darkness.

One man appeared and began caring for the kennels and the German Shephard hunting dogs. "They hunt in packs of eight to twelve dogs," he told me. "They are released from their kennels at dusk and return at first light. Each kennel has an electric switch to raise and lower the door."

When the dogs returned to the kennel, the concrete floors had already been cleaned with a high-pressure power washer and the waste flushed to the outside of the berm.

Once all the hunting dogs have returned and were inside their respective kennels, the doors closed, and the sergeant in charge began to feed the dogs.

He would take a half ration of raw beef, split the center and pour a couple ounces of black gunpowder in that split. He informed me that black gunpowder would keep them absolutely insane mean, make them aggressive. They would be half starved by dusk and want the taste of human blood, the taste for human flesh of the Viet Cong human body, and after all they are the evening meal. He said to me, "I hate this job, but it is much better than being out there," nodding his head, glaring out onto the high plane.

He informed me that a veteran had been brought over to care for all the dogs, as also a number of K-9 dogs that worked normally with an NCO on the K-9 patrols.

If one of the berm dogs came in shot or injured, the dog would be darted, put to sleep, then muzzled first thing. He then would climb down into the kennel with a 1911 .45 on his hip.

I asked about the handgun. He said, "You don't think I'm fool enough to go into a kennel unarmed with one of these savages, do you?"

At later times, on our greenline while pulling that duty with the aid of 7x35 field glasses, you could watch a pack of the dogs attack a two or three-man team of Viet Cong recon and watch the pack attack. The dogs would turn their heads in such a way to grab the Viet Cong in the throat after a good leap.

On occasion the Viet Cong would get off two or three shots, killing or wounding two or three dogs. Then you could hear the screams as the Viet Cong went down.

CHAPTER SIX

Captain Ben

Silvers was now gone, leaving just the enlisted men and Captain Ben. Soon Captain Ben would be transferred out to command a firebase for three months, then to II Field Base camp for six months. Then it would be stateside for him.

Four men from Duster Compound were being sent to Cu Chi. They were put on as Duster crewmembers due to the tremendous loss of troops killed in action. Many of the severely wounded troops with loss of limbs were referred to, in our slang, as basket cases.

We put two of the men in the back seat of the Jeep and one man each sitting on the wheel wells. This was a rare case, because I had to hook up the 1/4-ton trailer to carry their duffle bags and all of their personal gear.

Highway 13 was hot! We had to detour on sand roads, in some very remote areas.

We came up to a makeshift staging area with one Duster re-arming while three Dusters, one Quad 50 running in reverse and two 113's from the 25th were in an extremely heavy firefight with the Viet Cong.

We could see Viet Cong pop up from the vegetation. They were spraying the area with their AK-47's. They would occasionally pop up to fire an RPG. There were about six straw huts no more than one hundred feet from the Duster. The E-5 sergeant in the gun tub had the twin gun tubes pointing down the road. The M60 was pointed at and into the wood line where heavy elephant grass grew densely thick.

There were about twenty children who would run up to the Duster; one was two heads taller than the others and stayed in the center of the crowd. The sergeant swung the M60 into the direction of the crowd of children and screamed, "De-De!" Then they would retreat and run away, only to return and repeat this in only a few minutes.

Captain Ben had placed us in the ditch that was about eighteen inches' deep, spread apart one man about every six feet.

I was in the center of the enlisted men with Captain Ben to the far right. I had in my hand my M79 grenade launcher. The wood line had filled up. The last time the children came up all the way to the Duster, the tall boy, probably 14 years old, lobbed a M56 grenade toward the gun tub on the Duster. The sergeant in the gun tub squeezed off one round with the M60. The sergeant was on line 100 feet away while 12 Viet Cong approached, armed with AK-47's and one RPG pointed towards Captain Ben and his five enlisted men.

Captain Ben commanded, "Lock and load ladies, the shit is about to hit the fan!"

The one round from the M60 had killed the 14-year-old. However, that one round went through him and then through the stomach of an 8-year-old little girl. I watched as she tried to get up, struggling to do so, but could not.

All people from both sides froze. I looked across the Viet Cong firing line to the man on the far right, who was left handed, the RPG trained on him in the center.

It flashed through my mind in a split second that this child was so small. She was only doing what her mother and father had told her to do. She was not even old enough to know what was going to happen.

The sergeant screamed, "Medic, medic!"

The man loading 40mm striper clips into the ammunition wells, his legs severed at the knees, had blood shooting from the stubs every time his heart took a beat. About four not more than five times his body quivered, and he was gone.

The other wounded loader could be saved. Two men from the ammunition truck that sat in front of the Duster, about 12 feet away, rushed to put him on to the UHIC helicopter medical evac that would rush to the base camp hospital.

I threw off my steel pot helmet, threw the flack vest and then dropped my web gear. I came up and out of the ditch line with the 79 hanging in front of my left shoulder, a belt of grenades slung on my right shoulder.

Very quickly, after about four quick strides, I scooped the 8-year-old child into my arms looking into her eyes, and it was the look on her face that indicated, "I will live now that this American has picked me up."

I then looked up and down the Viet Cong firing line. They pointed their weapons straight up in the air and began to fade into the underbrush, one by one disappearing.

I cut around the front of the Duster and a few short strides up to the helicopter, "Take this child and save her!" I commanded, loudly.

"I'm not taking no communist gook," the medic replied.

"You will take her and save her or by God I will blow your asses out of the sky at 100 feet and send you all to Hell if you do not take her!" I demanded.

A young sergeant lying on a gurney, plasma bottle hanging with a lower abdomen wound popped up and screamed, "Take the kid, and get me to the hospital before I bleed to death."

The medic reluctantly took the tiny little girl, her feet dangling and already limp. He lowered her on to a gurney and went to work on her.

Wild eyed, I pulled out my eleven-inch blade, presented it rather wildly and screamed, "She better live, you hear me, or I will find all of you and cut your "balls" off. I have your bird number I will find you!"

The bird lifted off full throttle, screaming. I walked behind the Duster for cover, returned to the ditch line, sat down and gave a deep sigh.

At that instant, the commanding officer came from behind the Duster and screamed, "Name, rank, and serial number trooper. Your young ass will be court-martialed, and you will do a long stretch in Leavenworth for the stunt you just pulled!"

I looked up with a stare saying nothing.

The major screamed, "Damn it, name, rank and serial number!"

Captain Ben stepped in. "Give it to him, he has the vehicle number."

I replied "David J. McCormack, Specialist E-4 US 51951970."

The major red faced screamed, "You will be court-martialed, and do some years in Leavenworth prison!"

Captain Ben calmly said, "Not going to happen major."

"What do you mean captain, what are you saying," questioned the major.

"I will not let that happen, I will block anything you try to do to my trooper," Captain Ben replied.

"Captain, you will never make major," stated the major.

"Major, I don't give a shit about making major, I was drafted just like these kids, just happens I have a university math degree and was teaching high school math when I was drafted, went to officer's candidate school and here I am. When my two years are up I'm going back to my high school classroom where there is still a little sanity."

"Captain, maybe I will have you busted down instead, seeing how you are running interference for this rogue!"

"Major, I don't give a shit, I can be a private again and still be with these kids and be a part of them, which you are not capable of," answered Captain Ben.

The major just walked away.

Captain Ben then said to me, "Don't you ever, not ever, pull another stunt like that again. Next time I probably will not be able to protect you. They will hang your ass out to dry and there will be nothing I will be able to do for you."

A newbie sitting next to me took off his steel pot, popped me on the head with it and said, "You nut, you are lucky you weren't killed you crazy fool."

The rest of the troopers joined up on me, telling me never again.

I became quiet when after some thought I said, "Captain, I have to know that she lived; I must know."

"I'll check it out and for sure let you know in a few days," Captain Ben answered.

The four enlisted men were deployed by the captain. The captain stayed on firebase for three days to begin to learn about his new command he would be soon taking over. When the captain did return to Duster Compound via a Huey I asked him, "Did my little girl live?"

The captain replied, "Yes, indeed she did. Because you gave her back her life, she gave up her mother, her father, aunts, uncles, and cousins. Special forces, both U.S. and Vietnamese, went in the early hours of the morning and took every one of them out. She told the interrogators about the American soldier that had kindness in his eyes.

After absorbing what I had just heard, I asked, "What will happen to her now, no family left and all?"

"She will be sent to an orphanage far down in the delta where no one from her past will ever find her," captain replied.

"It's much the same as our orphanage here in Honi; she will be well cared for," the captain stated.

CHAPTER SEVEN

Specialist White

We had been having more claymores turned around at night, and more three-man Viet Cong recon teams probing the area. Then one evening after guard mount the CIA came.

It was explained a team would hit 2 Alpha. For some strange reason the posting and assigning of the positions was done early and Specialist White and I were assigned to 2 Alpha.

As we started for the hooch for weapons, web gear and other equipment, one of the young men from the radar room chased us down with tears in his eyes and choked up. He stated, "The CIA has a double agent that has lost clout with the other VC and he is going to lead an assault on 2 Alpha. The command will pull all of you on fire clearance, the officers and others will set up the second line of defense, the double agent will lead the VC right in to kill you three before you run for the northeast corner. The second line will kill all but the double agent, then let him escape so he will be good again. With the command of the VC you guys will die tonight!"

"Don't think so, just watch the kid tonight. Forget the club, and the evening movie. Watch me, it will be better," I said, feeling a rush of adrenaline.

Each man acquired his weapon and gear and went to the bunker.

Sure enough, one of the captains entered the bunker and said, "I'm pulling all fire clearance on this position until further notice," and he left. He started for the second line of defense.

The E-5 in charge started to bawl like a sick baby, "I'm going to die tonight, I'll never see my wife and baby again!"

I grabbed him by the shirt collar, slapped him three times across the face and screamed, "Shut up! Dry up you pantywaist pussy! You are no leader for this team; you cannot make a decision. You are surrendering command of this position to Specialist White or you will not last until dark, I'll slit your throat here and now boy!" I roared.

"Yes, yes, White you are in charge, I'll do whatever you say to stay alive to see my wife and baby again," the E-5 cried.

"White, hold down the fort here, I need to get to the hooch, pick up my extra battle rifle and get some gear," I said running.

"Got you covered little man," White agreed.

I started for the hooch, coming out of the bunker when the asshole captain popped up.

"Where the hell do you think you are going, specialist," the captain asked.

"I need to take a shit, sir." With a grimaced face, I added, "No toilet paper."

The captain said, "Take a shit in one of those artillery canisters from the blast wall, dump out the dirt and have a seat."

I replied, "I'm not going to let all you people see my hiney, not to mention my little buddy and best friend that all the ladies love so much. I need a little privacy in the latrine. Be back in ten."

The captain ordered, "Make it less!"

I hustled, went in the south door, straight through and up the hill into the hooch.

I grabbed my M16 for show. Grabbed my backpack and put in the three little section poles for the two-man tent. I then grabbed a small roll of 100 mile-an-hour tape. I grabbed a scrap of 3-foot square from a wool GI blanket. I put all the items in the backpack.

Hustling back to the bunker, I thought, "These 90-day wonder captains here have the intelligence of a sack of rocks; hope this dumb shit doesn't catch on as I carried the backpack back to the bunker."

"What the hell is in that pack," the captain asked.

"Extra toilet paper, sir." I answered. "I may have to go again after dark. No one can see my hiney then. Oh yes, and a couple of cans of beanie weenies for White and myself. If this E-5 wants any he has to go get his own; he is not in my platoon, so he has to furnish his own."

"Okay," the captain stated, and returned to the second line of defense. I took my M16 and went to work.

The E-5 asked, "What are you doing?"

I said, "Just watch and learn, pantywaist. I'm going to kill me another stupid gook tonight so you can go home to that wife and baby, just keep quiet."

I laid out my gear, taking the tent pole sections and taped them down. I staggered them around the barrel of my M16, protruding far beyond the muzzle. Then I rolled the blanket section around all and taped it down. I then pulled out a tubing cutter and trimmed about one third of the tip of one round off, then with the tip of the bayonet hollowed out the lead.

The E-5 asked, "Where did you learn all of this?"

"The round hollowed out and cut off, from wild dog hunting in Kentucky. The poles and blanket, from the MACV sergeant some time ago," I revealed. "The ARVN's use this trick at night against the VC. You can shoot at close range with no muzzle flash. Scares the hell out of Charlie," I said.

I then placed the more reliable AR-15 under the platform for the M2 .50 in case old "dumber than a sack of rocks," walked into the bunker.

I said, "Okay, Mr. E-5, place your M16 with the barrel sticking up in front of you and sit back so you have to stretch and reach for it. I'll do the same. White you do the same."

I continued, "We've got to fool this VC double agent. They are by no means stupid. The one that is stupid is the old sack of rocks, the captain."

Well, about 2130 hours the double agent and his team arrived.

He crawled in and would stand up slow and easy at about 20 yards. Us three had made a pact to pretend to be about half asleep, then pop up and make a quick grab for the AR-15's that were laying where old Charlie could see.

When they did this the Viet Cong would drop. This went on for about a half hour. I finally whispered, "It's time for me to kill this arrogant son-of-a-bitch commie. Put your hands down like the three of us have fallen asleep."

Old Charlie eased up, standing up.

In a blink of an eye I reached under the platform, pulled out the AR-15 and fired.

That Viet Cong had an astonished look on his face as if to say, "You cannot shoot me, the captain pulled your fire clearance."

In a short time his knees buckled, and he went down face first. The captain came running and ordered me out of the bunker, putting me at attention.

The captain slapped me as hard as possible across the face, knocking me to the ground. "You son of a bitch, you had to shoot him, get up now!"

He slapped me down repeatedly screaming the same. My eye glasses were broken.

The captain yelled, "What the hell do you have to say?"

"Well captain, with all due respect to what is on your collar, you have to shit sometime, sir." The meaning of that statement is this. I had hidden M56 grenades in the blast wall. If the captain went to the latrine with the cut off barrel under him, I would have lifted the rear door, dropped an M56 into the barrel, ran for two seconds, then dived to avoid the blast.

If he went and sat on a flusher with a closeable stall door this is what could happen. Back then there was a CV boot glue that hardened like today's epoxy glue. It would cure in maybe sixty seconds.

The captain sitting on the flusher with the stall door closed, could be slipped upon. The glue freely squeezed from the top of the outside of the door in 60 seconds would literally lock the man in the stall.

Walk to the exit of the latrine, cook the M56 one second, lob it close to the stall but not close where the captain could throw or kick it, and voila, one dead sack of rocks.

The lieutenant colonel had that captain cleared and off Duster Compound by 0900 hours, and transferred out the next morning. No one was ever allowed to know where.

After that night, the NCO and officers were deathly afraid of and feared me. I was "Billy the Kid."

CHAPTER EIGHT

Greenline

Every man on Duster Compound would have other tasks and duties other than his primary military occupational skill. One example of this is in the Deployment and Recovery platoon, where the strip team and the scrub crew were in some instances the same people.

Now their military occupational skills (MOS), were either wheel vehicle mechanics or track mechanics. The strip crew would remove good used components from the retrogrades. These were damaged tracks being prepared to ship to a contract shop in Guam for total rebuild. One operation was to cut the gun tubes off at about twelve inches at a time, placing the pieces on the scrap iron truck that would transport the scrap metal to the scrap yard via Long Binh.

The scrub crew had to scrub, clean, and sanitize the remaining human D.N.A. from the gun tube, driver's compartment and equipment with cut down concrete brooms, buckets of water, Tide laundry soap and plenty of bleach.

Now early on, when not doing Thunder runs to firebases, especially early hours while the strip crew would be stripping components and loading the Jeep, Chief Silvers assigned me to helping the scrub crew.

Two times straight on two mornings I just up-chucked. One can only imagine the smell after the three to five days of 98 to 102 degrees Fahrenheit with 100 percent humidity during the monsoon.

Well, the team leader informed Chief Silvers that I could not handle that duty well.

The chief sent the team leader to find and report to me. He said, "This thing of helping the scrub crew is just not working out. I cannot have you getting sick and then expect you to be 100 percent on your runs."

"How do you feel about handling claymores and grenades in replenishing the greenline each morning. You are still here on the base for what time the strip crew is still working to load the Jeep," the chief asked.

"Chief, I would love to have the chance and experience to do that job," I answered. "The more I could learn these things, the better job I could do while pulling greenline at night."

The chief agreed, "Good, the greenline crew it is for you. Report to the sergeant and Specialist White. White had been with the 25th until he was wounded and he is no doubt an expert in high explosive weapons."

I reported, the three-man team had worked the west wall from the gate down to 3 Bravo bunker. I spotted a claymore that had been turned around, and started for the claymore when the sergeant screamed, "Freeze! Don't move … Don't move!"

I froze, naturally.

The sergeant spoke to one of the team members, "Go to my hooch, and get the top to my zook suit."

The sergeant spoke to Specialist White. "Take him 100 feet back and tell him what he almost did. Orient him on the things he will need to know, and how not to get himself killed, and most important, how to not hurt or kill one of us."

The team member brought the top half of his explosives ordinance disposal armor suit.

"Here you go, Sarge," he said.

"What is going on?" I asked Specialist White.

"Just watch Sarge, then I will explain," White said.

The sergeant suited up, just the upper half of the suit, laid down and crawled up to the claymore. He took a four-inch paintbrush from his hip pocket of his jungle fatigues and began brushing away the loose dirt.

As the loose dirt was brushed away, an M56 U.S. grenade could be seen resting under the claymore that had been turned around.

The sergeant reached into another pocket and pulled out a common nail and a pair of pliers.

He pushed the nail, which was a little smaller than the grenade pin, through and holding the grenade with a death grip, bent the end of the nail with the pliers so there was no chance the handle could release.

White began to explain, "Old Charlie crawled in here last night, somehow beating the radar. He buried a short-fused grenade. Then he turned our claymore around and wired it down good. He hoped a newbie would cut it loose and kill or wound himself.

"From now on trooper, help White and do not touch anything unless me or White gives you the go ahead," the sergeant ordered. "This is the most dangerous job on this compound. This greenline is my responsibility, got that trooper?"

"Got it, sergeant," I replied.

The sergeant had one of the team members place the claymore and the grenade into the armored box.

After the greenline was once again "set up," as it was called, the sergeant and one team member took a deuce and a half out about 200 feet from the gate. The sergeant placed four ounces of C-4 on the items. He installed a blasting cap into the C-4. With a roll of wire and the aid of a detonator, he moved to the other side of the truck and detonated the grenade and claymore.

CHAPTER NINE

Whores vs. Prostitutes

I would continue to run Thunder Road. For the first six weeks after Chief Silvers put me on my own, I traveled to Cu Chi every day. I checked in and ran north beyond Tay Ninh, deep south down into the Mekong Delta, where Dusters need components for the machines.

I would stop about every two to three weeks to each of the four dedicated spies. I would give them the money then put the folded piece of paper in my boot. I would deliver it back to Chief Silvers. The Chief would read the material, take notes and then send me to the Special Forces side of the compound to hand it over to the MACV sergeant that had trained me, or I was to give it to the captain of the A Team.

Finally, one of the men in the platoon informed me that the little house next to the Special Forces side, next to the gate, was a little house of ill repute.

I said, "You mean those women who sit outside that little hooch are whores?"

"Hey man, they are not whores; they are prostitutes," the guy said, not really happy with my comment. "Whores give it away. Those are working ladies. They work to eat."

"One rule to work in that house is, they must have a dead husband who was a Tiger Ranger. Some of them have kids to feed, well for the most part they are on their own." He hesitated. "They must be tested by Dr. Soltis every month and even get physicals. These are our ladies; no visitors allowed from the outside," he said.

To hire one of the ladies, you got to use a pro-kit (to you civilians and armchair warriors, well it's a condom, short for prophylactic kit).

The man smiles and said, "Go up and try one out; who knows, you might get yourself killed tomorrow, so at least try to have some fun once in a while." As I turn to go he added, "Oh, yeah, one other thing, you catch anything off this compound, you know the drips or whatever, it's an automatic Article 15."

Well, the only thing I will share for the sake of this book is the little 78 pound Cambodian really took her work to heart and did the best she knew how.

CHAPTER TEN

Jinxie

Don Jenkins and I had tower with the little SP-4, Cookie; he worked in the kitchen. Don's nickname was Jinxie from the old cartoon in the 1950's Jinx the Cat. The three of us went up the tower at 1900 hours and relieved the man that had just completed his 24-hour duty.

Well, everything was going normal with the two-way radio: call MACV headquarters every hour on the hour for a communication check and on the field phone call our underground war room for the same, somewhere between 2300 hours and zero hour.

Cookie was screaming, "Help! Help! Man, please don't drop me!"

Jinxie screamed, "Mack get up! I just caught this little piece of shit smoking pot! Get his weapon and clear it! Get his web gear, bayonet, and everything!"

Jinxie, who was a very strong and powerful man, had Cookie by the shirt collar, and his hand behind the belt line holding him in midair outside the tower threatening to drop him to end his drug problem.

When I had cleared his weapon, and placed it all in the corner, Jinxie brought him in and sat him in the opposite corner. The little fella had his hands over his eyes, crying like a baby.

Jinxie said, "One of us is going to have to sit up and guard this piece of pothead shit. He could just kill us both. Chamber a round on your rifle and shoot the son of a bitch if he does more than blink."

Jinxie addressed the little fella saying, "You piece of shit, smoked up like you are, if the place were assaulted you could not even read the scale on the scope. You would probably get some of our people killed with friendly fire. This man here might just cut your balls off for a stunt like what you just pulled," Jinxie pointed at me.

Actually, Jinxie and I stayed up all night, about every thirty minutes taking turns. Together we tormented that little pothead until 0700 hours. Jinxie called the war room for a replacement for him. Jinxie had also searched him earlier and had taken the drugs off of him. When the officer of the day arrived with the Article 15, he made everyone on the base aware that you had better not get caught like that.

I never was put in the tower again on the colonel's orders. Really, he should have been court-martialed and done time in Leavenworth, but he wasn't.

A couple days later, Jinxie was in hot water. We had a little club, an EM club if you could call it that, which consisted of a bunkered building. The interior was about 24 feet wide by about 30 feet long. There were no windows, and the exterior on the outside wall were nothing more than sand bags stacked about two feet thick. The walls had heavy poles to support about three feet of thickness on the roof. There were two or three six-inch pipes for vents through the roof with kind of cones to allow some of the tobacco smoke out.

There on the end was a 90-degree kind of room or corridor to enter that was only sand bags for blast protection. One had to enter a screen door into the main club. The ceiling and plywood walls had a charred finish plus clear varnish. It didn't look too bad inside; the floors were shipping pallets like the mess area. For tables, there were big empty reels that wire came on. The reels created a four to five-foot table once they were turned on one side. About four or five feet across straw bottom chairs bought at Honi Village also served us well.

There was no electricity. There was a vintage Coca Cola ice chest that was dated back in the day. There was a very beautiful old French bar with the red cooler sitting next to it. The sergeant first class in charge of the enlisted men used his own money to run the club. There were big candles and a couple of oil lamps that no doubt were bought from one of the many stands from Honi Village. There were never more than a dozen enlisted men in the club at one time as everyone else had greenline or half a dozen other duty commitments; wrecker drivers, two of the three, gone bringing in retrogrades and other assignments. At least one of the three traveling mechanics were always gone changing a Duster engine on some firebase.

I had not had the time to be at the club but at last, a free evening, with no greenline or anything. Jinxie and Big Shook, a wrecker driver, asked me to go to the club for a beer or two. Jinxie was a mechanic, one of three traveling mechanics, and had just been dropped off late that morning, from a UH-1D helicopter after being on firebase to change an engine on a Duster. Now the rule was after changing an engine, the mechanic had to stay with the gun crew for 72 hours to be sure the machine was 100 percent and would not be down on road convoy or on a sweep or whatever.

When these traveling mechanics came in they usually hadn't showered shaved or whatever. Of course, Jinxie always carried a toothbrush in his blouse pocket. Now when I would take Jinxie out to change an engine, he took his empty duffle bag. We would stop at the 199th Light Infantry Brigade PX. He would fill the duffle bag with beer and usually two fifths of bourbon.

The gunnery crews would always give back what he spent and Jinxie would stay a little tipsy while working, changing the engine during the hours of daylight, and pour it down after sundown.

Well at noon, Jinxie was in the hooch, freshly showered and ready to take his 24-hour stand down. He planned to get a little sleep and then drink like it was going out of style.

Jinxie asked me if I had any booze and I replied, "Just this open bottle of 90 proof Smirnoff."

"Can you let me have some?" he asked.

"Sure, how much do you want?" I asked.

"Just a half canteen cup will do this evening until the club opens," he said.

He gave me like a quarter or fifty cents. When Jinxie got something, he would always insist on paying for it. All three of us had thrown down a plate of the little more than garbage from the mess hall. I had quickly showered, shaved, and whatever.

This was the first time all of us had time to be together at the same time. We went into the club, ordered a beer each, and relaxed. Now Jinxie was already about two sheets to the wind from the Vodka I let him have earlier. He sipped it through the evening.

The more Jinxie drank, the harder he got. He pulled out his wallet and showed us a photo of him sitting on a wild and fine chopper motorcycle from about two years earlier; it was a 750, or it was before he modified it.

It had a heavy rake on the front. Monkey handlebars, custom exhaust system with the pipes turned up in the rear. A seat that was oversized. He began to tell us about the oversize cylinder jugs with oversize pistons, high compression, the radical racing camshaft, and aftermarket carburetors. He then told us what she ran in the 1/8-mile back home. It was super-fast, even for a bike.

Well, then he started going on about the Watt's Riots of '65, and how he and two other bikers from his local town and from their motorcycle gang had rode the fifty miles or so into Kansas City to take in a movie at a theater.

Jinxie then pulled up both trouser legs and showed us scars up and down his legs from surgery where both legs had been broken in '68. He then pulled his blouse open and showed scars from surgery where three ribs were broken and had to have steel staples to repair. He then put his hand on his head and spoke of a skull fracture. He was now three sheets drunk, stood up, and said, "Them damned n----- left me and my biker buddies for dead on the streets of Kansas City back then."

He then proceeded to make more racial remarks, when the African American sergeant first class pulled from under the counter and presented a weapon. It was an M1A1 GI Carbine with the stock cut off; the barrel cut to about ten inches with front sight reattached, and the sling remounted.

The sergeant screamed, "Get the hell out of my club you redneck honky!"

Jinxie blurted out, "Throw down your guns, knives and tire irons and come out and fight me like a real man, you n-----."

Needless to say, when that M1A1 came from under the counter, Big Shook and myself unassed those straw bottom chairs. He placed his back and big ass against the side wall about twelve feet to Don Jenkins' left. You can bet your military retirement that my little ass was right next to Big Shook. We were prepared to make a new door, so fast anyone's head would spin.

At that instant, the sergeant major stepped in the door, arms crossed holding both pistols still in the custom shoulder holsters.

He screamed, "Sergeant, I need your total undivided attention, this fool is unarmed, so drunk he is no threat to you. Put the weapon back where it came from. I'll handle this fool!"

Big Shook and I were both shaking, sure that Jinxie was already dead.

The sergeant major dragged his drunk ass outside in the open air and said, "Boy, do you want to work for me or the lieutenant colonel. If you work for him your sorry ass will wind up in Leavenworth making gravel from big rocks with a sledge hammer!"

Jinxie, so drunk he could hardly stand, started laughing abruptly.

The sergeant major asked, "What the hell do you find so funny son?"

"I guess I showed his chicken shit ass who wears the balls around this compound," Jinxie ranted.

"I'll ask you again, trooper, who are you going to work for myself or the lieutenant colonel."

Jenkins replied with a laughing slur, "Sergeant Major, I am going to work for you; I hate officers!"

"You two, get this piece of shit up the hill to the showers, throw him in clothes and all. When you think, he is sober enough, carry his dumb shit ass and throw him in his rack." Pausing the sergeant major added, "Have his sorry ass in front of my desk at 0700 hours. He is permanently barred from your club and is not to be within fifty feet of the club and he is not to be any closer than 50 feet of the sergeant, understood?"

We carried Jinxie to the shower, laid him on the floor, and turned on the cold water. He just laid there, laughing and every few minutes would say, "I love you guys," and continue laughing.

After about thirty minutes we helped him to the hooch and threw him on his cot.

Big Shook rounded up four of five guys in the hooch and announced, "Jenkins is in deep shit; we've got to work together and keep him here in the hooch until he is sober. In the morning, he will face the sergeant major."

Everyone agreed. We all pulled one hour shifts until about 0400 hours and then got a little shuteye.

At about 0645 hours the next morning, Big Shook and I escorted Jinxie to the sergeant major's office.

Chief Silvers was present. Just outside the door Jinxie said, "Boy, I am in deep shit; the chief is here."

Everything the sergeant major said was repeated.

The sergeant major turned to Chief Silvers, "I don't want Jenkins on this compound; therefore, you keep him on Duster engine swaps in the bush. When he comes in for his 24 hour stand down, I want his dumb ass gone again for the next one."

Chief Silvers said, "Understand fully, sergeant major."

The chief took the three of us to motor service. He told us to begin our duties. He took Jenkins into his office, and ordered to Russ, "I'm not to be disturbed until I'm through with this one."

Russ replied, "Sir, yes sir!"

The chief started in on Jenkins, telling him one more screw up…; well, the chief made it clear he would be pounding big rocks for a good twenty years.

CHAPTER ELEVEN

Searchlight Team

Three African Americans who were on a searchlight team came in to stand down for eight hours. It would be normal when the colonel gave permission for daylight hours, as the infantry company would put two grunts in their tower. This was done during lull times to give the team time to sleep.

You see when things are hot, the team would be up all night working the infrared and white light for Dusters, Quads and all other people regardless of what weapons. The word was these three men had blown a joint out on the berm. I cannot say if they did or did not; you know how things go like that. Well, a couple of white guys claimed they could smell the weed on them, one thing led to another, and in an instant, racial names were flowing from both sides with no limits, weapons were pulled, locked and loaded safeties in fire position. The weapons included one M79 loaded with a high explosive grenade, and one M60 with a 100-round belt ready to fire.

The sergeant major double timed, stepped between the two groups. His 45s were locked, loaded and ready to fire, one in each hand pointing a pistol at each group.

Standing in the center, the sergeant major screamed, "Clear those weapons now!" With the safeties in fire position, trigger fingers starting on the trigger he screamed, "I'm giving you people to the count of three; at three, I am going to start firing and put all of you down, and anyone still alive will get one round in the head!"

"One…two," all weapons were immediately cleared and placed on the ground.

"You three go sit on the east berm!" nodding at the three African Americans.

"You five go on the west berm; anyone, I mean anyone moves out of position, you are dead!"

In the meantime, as the crap started, the colonel with his M16 and two of the captains with 45s – two men each were covering these eight men ready to annihilate them if things went south. The sergeant major, both pistols in hand with safety still in fire position, was ready to talk and everyone listened.

"The word is if this shit starts again, the colonel's staff and all officers available are going to fire everything we've got into all of you. There will not be much left to put in a body bag, you assholes got that," he shouted.

The answer was, "Yes, sergeant major, sir!"

Silvers commanded all of us in his platoon to stay out of that mess.

A couple of days later another mess, two of the searchlight team members were arrested for stealing a Jeep belonging to a Marine Corp major. They stole it from the loading docks in Saigon.

The field commander of the Corps was screaming for a general court-martial. He wanted to send the two 19-year-old boys to Leavenworth for twenty years.

The battalion commander, the lieutenant colonel and the sergeant major went to MACV and spoke with command. They made a secret deal that the Corps knew nothing about. They were brought back to Duster Compound. Their heels were locked in front of the lieutenant colonel. The sergeant major, one 29th Battery commander, and the fire sergeant were present. The lieutenant colonel began to speak.

"You have got to be the dumbest assholes in this man's Army. You stole a Jeep from a Marine Corps officer, you repainted that machine and put the I-29th on it." Shaking his head, "The MP's informed me they caught you two by the serial number on the data plate." After a long deep sigh, "When you steal a piece of equipment like that, change the data plate!"

"Well sir, you know we had turned our Jeep in as a combat loss, and there wasn't a data plate left, you remember, a B-122 gook rocket destroyed our machine on the firebase," one teen spoke up.

"We were just tired of having to hitch rides to come in, and if we have to move the searchlight to another base we'd have to bum transportation, sir," the other teen added.

The lieutenant colonel said, "Well, the smart thing would have been to remove the data plate, come into Duster, and change to another Jeep before going through an MP check point at the base camp."

"You two have outstanding records to this point. You both have one Purple Heart award and Commendation Metals for valor. I'll see to it, you can forget the Good Conduct Award. You both are clearing my compound, today. You will be taken this evening to the 90th

Replacement. I have arranged for you two to catch a flight out of Binh Hoi tomorrow as people at MACV have bumped two enlisted men so you can have their seats.

"I have taken what leave time you have left. I have already contacted a lieutenant colonel in Germany. You both are being sent to his battalion. You will be on a searchlight there waiting for a communist Ivan the Terrible to cross the border from East Germany. Don't even ask for a weekend pass or anything when you arrive.

"When not on the searchlight you will be restricted to the post. Go AWOL and I will personally see you troopers get that twenty years in Leavenworth. Now take a clear blouse, cut your name off, put your dog tags in the pocket. You stay to yourselves at the 90th. Do not leave their area. Talk to no one. I have put myself and the sergeant major way out on a limb for you.

"Captain, escort these two to clear my compound, I'll assign one of you NCOs to stay at the 90th and see them to the plane and see they board the plane," the lieutenant colonel ordered.

"By the time the Corps knows you are gone you will be in Germany. I don't think they will ever find you, by following my instructions you will ETS from Germany. If they happen to come here, we will say we guess they went AWOL and of course they will be looking for you in this country. Now get the hell off my compound, and good luck in life. I'll expect a letter with no return address from Germany from one of you. Now go!"

According to the lieutenant colonel's clerk, he received a letter in about a month from them; all was well in Germany.

CHAPTER TWELVE

Monkey

I went on making my runs pulling my greenline about every other night, catching first shift in place of the E-5's so I could catch about five to six hours of sleep.

It was no more than two days later that the chief called me in and said, "You need to transport Jenkins to Xuan Loc. He will fly out of there up further north off Highway 1; you know I'll never allow you north on 1.

"You and Jenkins need to leave in one hour. Get your gear and remember to take your poncho roll."

I said, "Sir, yes sir, and thank you sir!"

The chief asked, "Thank you for what, Mack?"

"For not pushing to bust Jenkins and just letting the sergeant major handle the problem with him," I answered.

The chief replied, "Jenkins does not conform to military life like the rest of my platoon, Mack."

I said, "Chief, Jenkins and I caught little Cookie smoking weed in the tower a little while back and Jenkins handled the situation much better than if he had not been there with me. He hates dope with a passion."

The chief said, "I know, but that does not excuse his many faults."

"I know sir, but Big Shook got the platoon together, and, well we are going to keep him straight if he has to have his ass beat, and if that happens, it will take three of us to do that, sir."

"Good, you guys do whatever is necessary. Maybe a good ass whipping will make him conform."

The chief added, "Go now and prepare early, double check your gear; I can't afford to lose any of my platoon, especially you. I am making plans for your future after this mess is over."

"Sir, yes sir, chief," I replied.

Jinxie and I rolled out on time. He had the usual for his mission — toothbrush in his blouse pocket, his toolbox and his empty duffle bag.

We did the usual. We stopped at the 199th Post Exchange. He went inside, brought out cases of beer and two fifths of bourbon and, as usual, filled the duffle bag.

As I started the Jeep, two Green Berets pulled in driving a 3/4-truck. Jinxie threw his left hand into my chest and loudly said, "Wait, I know these guys. You are going to love this, just sit and watch for a few minutes."

As the two Rangers exited the truck, one of the sergeants said, "One of us has to stay with him, you know how he is," referring to a little spider monkey, dressed in a little custom vest. It was Army green with a full size 5 Special Forces patch on the back and also at the bottom on the front. He had his little green beret on his little head with a black elastic strap under his chin. He had a bright red dog collar around his little neck and about eight feet of dog chain anchored to the right-side panel of the truck.

The monkey was eyeing us as we sat; his little legs were crossed just like a human and his little hands clasped behind his little head. He, no doubt, was imitating what he had seen of these Green Beret Rangers. The second sergeant said, "He will be okay where is he going to go. He is chained down, come on in with me."

The first Ranger agreed. "Okay, okay, we will only be a few minutes, but I am still a little worried."

They both started for the PX door. Just as they went through the door, the little monkey sprang up as if he had been hit with a bolt of lightning. He ran and jumped up, glaring through the rear window opening, his feet clenched on the top rim of the truck bed. He had his little hands holding onto the bottom of the window frame. He turned loose with his tiny little hands and pushed the tiny little beret back and forth on his little head, chattering and making those monkey sounds very excited.

When the Rangers totally disappeared, he leaped after spinning around. He ran to a military food cooler sitting there that set directly in front of him when he was kicked back sitting on a folded tarp with his legs crossed and his hands clasped behind his head.

He made a couple of leaps, jumped and landed, buttocks down on top of the cooler. He reached down and grabbed the first of the four latches keeping the lid of the cooler closed. He grabbed the first latch and pulled and struggled with both little hands. You could see him getting red faced, then all of a sudden, the latch popped free. He bounced over to the second latch, and then the third and then the fourth. The lid could now be popped open.

He jumped down, feet on the truck floor, grabbed the lid with both little hands, and strained with all of his strength and then some. The cooler top popped open and fell with a flop and thump.

He dove into the cooler head first, with only these tiny feet and buttocks sticking out of the cooler. He popped back up with a can of beer in both hands, ran around the cooler, and flopped down in a sitting position with the can of beer between his little legs.

With both hands, he clasped and pulled on the pull tab. He kind of turned his little head towards Jinxie and myself, straining on the pull tab. We could hear the tab give way.

He turned that can up and proceeded to drink, his tiny throat bounced up and down as he continued to pour down that entire can of beer. The little throat stopped bouncing up and down — the can was empty!

He pulled the can down, peered into the opening with one eye, and threw the empty can to the other side of the truck, then crossed his legs, and clasped his little hands behind his head. In about one minute, he let out a big burp.

In about one more minute he kind of leaned to one side just like a human, then farted.

Jinxie and I were laughing so hard my ribs were in pain.

About then the two Rangers exited the post exchange with four cases of beer each and a fifth of fine Kentucky bourbon laying on top of the beer.

The one Ranger who had been reluctant to leave their monkey alone screamed, "I told you so, he's drunk again!"

As Jinxie and I drove away I asked him, "How do you reason that little creature could hold so much methane and still be alive." He laughed so hard that I wondered if he was going to split one of those stapled ribs of his.

We headed east, a typical run for the both of us.

CHAPTER THIRTEEN

Wolf

About 15 miles east of the 199th Light Infantry Brigade base camp, we fell in behind a slower moving water truck with trailer. You could tell it was fully loaded.

We kept a safe distance from the truck, about 100 feet back. About that time, we were approaching a blind hill and no more than 200 feet from it. A government Jeep painted grey came wide open behind us, swung into the oncoming lane around us and the water truck, and came head on into one of the very old French garbage trucks. They just almost made it around the water truck… "almost." The left front wheel of the garbage truck went right over the Jeep and the driver looked like an opossum hit by a Mack truck. The passenger in the front was severely injured, but the one female in the rear seat was launched about 15 feet in the air. She landed with her back hitting in tall elephant grass, and got up uninjured.

About this time, the metal parts blown from that old M38 Jeep hit the ground. Another M38 stopped behind us. It had a full bird colonel, a major and a command sergeant major. A young captain was driving for that ARVN colonel. The colonel walked quickly to the passenger side of my Jeep and in very good English asked, "May I please use your radio?"

I handed him the radio, "Please do, sir! Get that one man a medical evac as soon as possible."

The captain cranked the two channel dials to his shown Saigon group. In no more than five minutes a medical Delta helicopter arrived along with a gunship.

The gunship sat and hovered about four feet off the ground about twenty feet to my left. The door gunner had his door gun aimed over our heads pointing at a rubber tree forest about 200 feet to our right.

In no more than three or four minutes Jinxie shouted, "Shit man, there are people moving around in those trees!"

At about that instant the water truck driver, battle rifle in hand, double timed it to my side of the Jeep saying, "We got people in those trees, fellas."

The truck driver then said to Jinxie, "Hey man, better come around to this side of this Jeep, the shit could hit the fan anytime."

We all three took cover on the left side of my Jeep.

About this time the medical evac lifted off. The water truck driver said, "I'm making a run for my truck, cover me in case some shit starts."

He made his mad dash for the short distance he had to go.

"Come on Jinxie, let's get rolling before something bad takes place," I shouted.

The door gunner threw a thumb up, and I started the Jeep engine, started out with the clutch, threw him a return thumbs up with the left hand. He nodded his head, then the Huey lifted off.

I was nailing the throttle some. Just before we topped the hill, a shaped heat rocket went into the gas tank on the ARVN's Jeep. At the same instant two RPG high explosives hit them at the same time.

It brought the Jeep about six feet off the ground. Four bodies launched about twelve feet into the air, burning as if they were covered with napalm. The reality was they were cooked, in raw gasoline.

About eight or so Soviet AK-47s were spraying them down; talk about an absolute overkill, it sure was. I saw everything in those huge rearview Jeep mirrors.

I slammed on the brakes while the Jeep was still in low gear. We stopped about 20 feet from the crest of the hill. I reached around for my M79 and was looking straight to my right and a Viet Cong was running with his RPG, came to a stop behind the tree starting to line up his sights on us.

At that instant Jinxie threw his left arm around my neck and screamed, "Drive or I blow your brains out and say Charlie got you!"

We moved about twenty-feet and the high explosive RPG rocket struck the hill twenty feet behind us, about thirty-five feet away from the road to our left. Two fragments tore through the canvas on the right rear of the Jeep. Jinxie would continually tell me I was crazy.

About five miles down the road, I told Jinxie to get that muzzle out of my ear and away from my head. I said, "If I hit a pothole at 60 miles an hour and that rifle discharges, you will die right along with me." Jinxie removed the rifle from my ear.

About a half mile from Xuan Loc, I picked up the radio microphone and said "Xuan Loc, Xuan Loc, Duster Chief calling, pull the wire, pull the wire, I'm coming in."

The reply was, "Go back, go back, we are hot, we are hot!"

I replied, "Damn, it's Billy the Kid, I say again, Billy the Kid, pull the damn wire or I'm coming over it!"

A staff sergeant, CAR-15 in hand, came running to us crouched down, flack vest on, steel pot on. "Where the hell have you guys been, we thought you guys were dead. They got bodies on the way to Long Binh. They thought it was you two, and had picked up a couple of hitch hikers."

I said, "Some kind of mistake."

Jinxie yelled out, "This bastard is some kind of crazy!"

The sergeant said, "You guys got to help us, shit we got better than a battalion of Viet Cong walking in. We are going to be overrun! We need every swinging dick and then some, follow me."

I drove the Jeep all the way in. I shut her down five feet from the fourteen-inch-high excuse for a berm. Jenkins was hunkered down as low as he could, I screamed, "Come on, get out and help me kill some gooks."

We flopped down in a prone position, put his face down and began to burn full auto battle rifle at about 30 degrees up, taking moon shots hitting nothing.

I started popping off rounds. Charlie would just miss a step with the first round that would strike main mass center (upper center of the chest) but did not go down. One more round to one pelvic, one to the other pelvic, most went down with this. That is what Silvers had taught me. The Viet Cong were no more than fifty yards out.

I was firing a round every 1.5 seconds or so. I popped one main mass center face. When that 5.56 round hits the head, you don't need to know what that looks like. But it's a hell of a mess.

There were a small number of us, with two M60s, two kids throwing grenades about every 1.5 seconds each. The remainder of this little fire team was raking the Viet Cong with M56's, and a total of four troop's running back and forth bringing up cases of M16 grenades and mini cans for the M60 machine guns. I was running low on magazines, so I kicked Jinxie and screamed, "More ammo!"

His feet hit the ground about twice and in no more than three seconds he threw me two bandoliers from my mini-can in the Jeep. I was putting a round into them every 1.5 seconds or so. I wasn't counting rounds, but I was staying calm enough to count my kills — exactly 87 kills for me in no more than 15 minutes.

Charlie backed off, so we all quit firing and allowed the survivors to collect their dead. Some of them put them on the bamboo ladders and took their dead with them.

I'll admit today that I had a way of converting total absolute fear into a rage you just would not believe. There was a lust for the blood of these people that were killing my friends, my brothers, my countrymen. I could at times believe I could taste their blood as a wolf does its kill.

It was earlier in the day when Jenkins and I left. I had two firebases to hit dropping off breechblocks and other components that day, and had nothing to eat for a noon meal.

I rolled back into Duster, and as I cleared the gate at somewhere about 5:30 P.M. civilian time the gate guard screamed, "Hey kid, you are still alive!"

I replied, "I hope like hell so, unless this is Hell."

When I rolled in, Russ came running from the office trying to give a hug and crying and said, "We thought you were dead, they got bodies at Long Binh."

I pushed Russ's arms away and took a step back.

I said, "I'm fine, don't touch me! If I buy the farm, go and kill three of them and call it even for me!"

Russ started to say something and the chief stepped up and interrupted saying, "Leave him alone Russ, let him deal with this his way. You people get back to work. Do not bother him!"

I took my grenades to the bunker to store them. I just sat there and stared out on the high plane, somewhat nauseous, looking down at my hands as they started to shake.

At about 1800 hours, the end of the normal work day on the compound, the chief stepped into the bunker and said, "You were up for greenline tonight, but I put one of the men in your place. Have a few drinks of the Smirnoff, get some sleep and rest. It's another day tomorrow, you've got to make it back to the World and put this insane thing behind you."

The next Sunday morning after this bad day, I washed my Jeep, then changed the oil, lubed it and checked all CV joints for looseness. I covered anything that could leave me stranded in the middle of Viet Cong territory. Having finished everything that was to be done that morning, I just sat in the doorway facing south, looking at the south dirt wall and enjoying the peace and quiet after a very hard grueling week.

Out of the quietness, mortars began to fall, many of them. I instantly dropped the Pall Mall cigarette from my left hand. I dropped down, low crawling to the maintenance pit about 12 feet inside the building, going head first into the pit.

I backed into the corner, sat down and fired up another cigarette, hand shaking, using both hands on the Zippo. When it was all quiet again Russ was counting heads for the chief, and kept calling "Mack, Mack!"

The chief called out, and I answered, "Here in the pit, chief."

One of the men called out, "You can come out now, it's over man!"

The chief barked out, "He can stay in that pit as long as he wants, all of you just leave him alone!"

CHAPTER FOURTEEN

9.5 Jungle Boots

Three of us were resetting the greenline and replenishing ammunition in the bunkers. Specialist White drove the old deuce and a half. I was not allowed to ride in the rear with the grenades and claymore mines.

I was on the right side standing on the running board. Someone had replaced the bolts with standard 3/8" bolts. The 9/16" wrench fits that particular bolt.

I must have been issued the last two pairs of size 9 1/2 regular jungle boots with the older, wider cut outs in the heels. The cutout in the heel of the boots continued to go down over the bolt head, and I would have to pull up to free the foot. The old battery box door was warped at the bottom to the outside and away from the center of the truck. The outside of my boot sole would slide under the door, causing me problems.

We were about halfway through resetting the greenline at approximately 0830 hours and returning from our underground ammunition dump with fresh grenades for the M79, two cases actually, and had refilled the claymore vest on one side.

We went 200 yards south on the berm road and turned east. Just as the specialist turned to the left and just started east, it began raining 81mm mortar shells.

White made a hard left on the center road that ran north and south in the center of the compound. Some Viet Cong was bombing our truck and walking the mortar shells in on the truck.

It flashed through my mind about two more seconds and a mortar will be in the truck bed, detonate these two cases of grenades and there will not be anything to send home in a body bag.

The truck was traveling at a good twenty miles per hour or so.

"It's time to jump, and hit the ground running," I thought. So, I did, except that my left boot was hung. Down I went face first, and landed on shattered sea shells. My face was cut up with superficial cuts, but one piece cut well into flesh, almost through in the center of my little mustache.

Pieces were driven into my left arm; one arrow shaped piece was lacking 1/8 inch going all the way through my left arm. I was being dragged on the ground, left foot still flat on the running board and being dragged in front of two sets of dual rear tires.

I realized instantly that if that boot turned loose I would be crushed. I knew that the medics would have to scoop my remains up with a scoop shovel.

Twisting my body to attempt to break free and roll out from those dual wheels was just instinct. About the fourth try it worked. However, that turned the left foot and leg from the ankle into the hip at least ninety degrees more.

I was on the ground trying to get up and stand. Blood was pouring out of my arm injury just below the elbow. Everyone on the compound was in air raid shelters.

One man peering out at me just looked and froze up when I cried out "Help me, help me, I can't get up!"

Specialist White had stopped the truck at the dead end of the sand road, bailed out and came around the front of the truck. He screamed, "Man, if I had known I would have stopped." Mortars were still falling and detonating. White, the only man with balls that day, ran and picked me up.

White took my bleeding left arm, put it around his neck and with his left hand held me by my wrist, with his right hand he took my waistband. Walking on my right leg by myself with White's "uncommon valor" got me to the dispensary where medics and the battalion surgeon, Dr. Soltis, went to work with utmost speed.

A Latino medic commented, "Mackie you be okay, you see."

The medics had to work fast to save me. There was so much blood loss.

One medic on my right slapped the needle in my right arm, put on tape, and hung the plasma bottle. On his left, the Latino medic working feverishly pulled out the long arrow shaped stainless sutures.

Specialist White was by my side, holding my right hand clasped in both of his hands. White reassured in a soft-spoken voice, "You are going to be okay. We've got the best surgeon and the best medics in this whole war."

When I started to fade Dr. Soltis barked, "Elevate his feet, treat for shock, get some ammonia under his nose!"

I came back to full consciousness. My head raised and my right hand pulled out of White's. I said, "Get that stuff out of my face!" I pushed the medic's hand away.

"Before we do anything with this foot and leg, we are going to give you an inspection," said Dr. Soltis.

"No morphine, Doc," I said.

"We are going to use codeine," Dr. Soltis confirmed. The medics removed my left boot as gently as possible.

The doctor examined my foot, "The bones are broken; we must send you to the hospital for x-rays and to set the bones."

"Nah, Doc, the ankle is just sprained, patch it up. I've got a greenline to set and a Thunder run up north. A crew split breechblocks last night. If they don't get them, a lot of people are going to die or really get screwed up."

The doctor replied, "Okay, let me see what we can do."

They took this medical device to immobilize my foot; it's a kind of plastic that the foot goes into, and is taped down without a boot of course.

The medics with scalpels whittled it down until it would fit into the jungle boot.

They put it on my foot and taped it down really good, then allowed me to put my boot back on. So, everything had been treated.

Soon a second bottle of plasma was hooked up. When it was empty Dr. Soltis stated, "We are going to send you to the hospital at Long Binh for whole blood. You are still a pint low, and we have given you two plasmas, and that is the limit."

Now Silvers had told me not to take whole blood, as some of it wasn't screened and a patient could wind up with hepatitis.

I replied, "No whole blood for me, Doc."

"You have to," Dr. Soltis replied.

"No, it's the church, see, Church of God," I answered.

Dr. Soltis looked at the dog tag and commented, "Okay, but any further problem you will have to get more blood."

I commented, "I've still have a greenline to finish and at ten hundred hours I start on my Thunder runs."

The doctor reluctantly released me. I walked out the door, made it about 20 feet and passed out. My blood pressure was too low.

The medics carried me back in and placed me back on the gurney.

The doctor handed me a cup of liquid. I asked, "What is that?"

"Alcohol mixed with water and Tang," the doctor answered. "It will help raise your blood pressure; it's too low."

As he prepared to leave, putting the left foot on the floor, pain shot all through the foot and all the way to my spine.

THE FOLLOWING PHOTOS ARE FROM THE COLLECTION OF DAVID MCCORMACK.

Big Shook's rig on Duster Compound.

Bill Timolsic welding with oxygen/acetylene, making repairs on Duster Compound.

David J. McCormack on Duster Compound.

Specialist White saved my life during a mortar attack.

David McCormack with souped up M151 Jeep.

David McCormack with the Iron Horse Duster Chief.

David McCormack guarding a UH1D.

Quad 50 'Top's Popcorn Machine covered the entrance gate.

David McCormack and gunnery crewmember on Top's Popcorn Machine.

Billy the Kid with his Iron Horse heading to a firebase.

David and Duster Chief near the greenline perimeter.

"Rollin Cobra" on a base camp

David McCormack

Southeast end of Duster Compound.

Local Vietnamese lady washing clothes for gunnery crewmembers.

Donations from enlisted men built God's house for Sunday evening worship, their only time off.

David "Billy the Kid" returning from a Thunder Road run.

Above and next page: China Beach R&R.

David McCormack and Big Shook having a photo op.

Local lady who washed clothes at base camp.

Don Jenkins (Jinxie) having a midnight Christmas Eve nightcap.

David (right), Don Jenkins (center), and Sgt. Bill Timolsic, Christmas 1970.

Base camp's Charlie Brown tree, Christmas 1970.

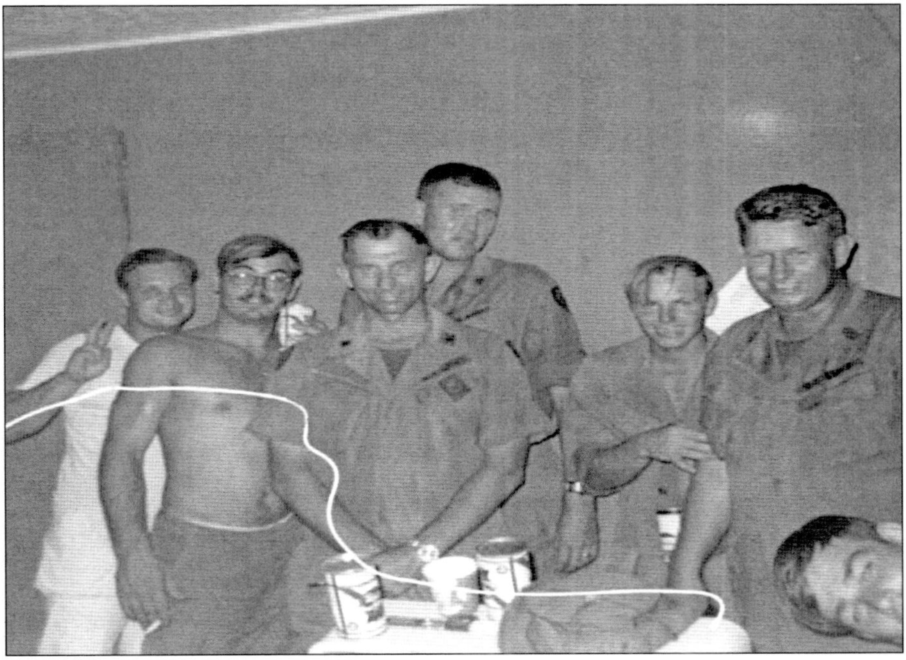

Mechanics from H&H 5/2, Christmas Eve 1970.

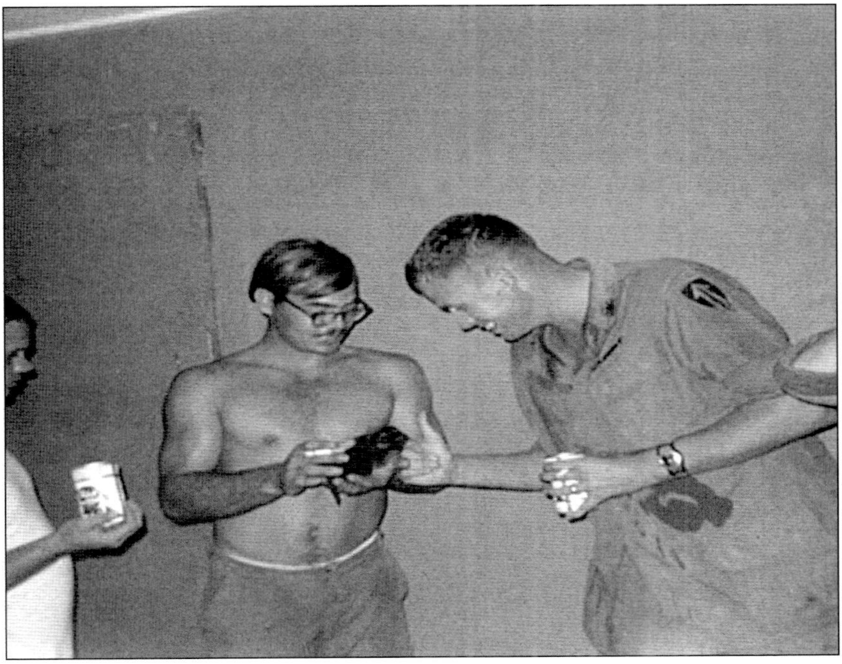

Captain Moore, Don Jenkins (center) and Cook (left), Christmas Eve 1970.

Bob Hope entertaining the troops Christmas Day 1970. David and friends were hungover.

David taking a smoke break in a no smoking area.

David with an M2 from a Quad 50, Base Camp, March 1970.

Black Virgin Mountain in Tay Ninh Province.

Jungle view from a Quad 50.

Fire Support Base – to the right is the 29th Watch Tower with infrared over white searchlight.

Old 199th Base Camp taken over by the 25th Infantry Division. Pictured is a UH-1D Huey.

Duster Compound Delta Battery, a Duster leaving to go to a firebase to fight communists.

Duster Compound hamlet and A-Camp, 24/7 manned watchtower in the distance.

Delta Battery Duster

A Quad 50 gunnery crew D 71st heading to a firebase.

Duster on Duster Compound.

Tran Van Loc (Charlie) and his new bride in Honi Village.

The wedding party in Honi Village.

Children attending the Catholic wedding ceremony.

Two guests at Tran Von Loc's wedding thought to be double agents; Sgt. Bill Timolsic in the background.

THE FOLLOWING PHOTOS WERE TAKEN BY SFC E7 (RET) ROBERT W. "GRIF" GRIFFIN, WHO WAS A COMBAT PHOTOGRAPHER IN VIETNAM FROM 1966-68 AND 1970 WITH THE 1ST CAVALRY DIVISION AIRMOBILE. ALL PHOTOS USED WITH PERMISSION.

Robert W. Griffin "Grif" in 1966 when first assigned as a combat photographer.

SFC E7 Robert Griffin checking out locals in Dak To, 1966.

Vietnamese kids raise their flag early in the morning at An Khe, 1966.

CH 47 Chinook takes off at An Khe airstrip, 1966.

54th Transportation delivering needed supplies to build barracks in 1966.

South Vietnamese fort in the Central Highlands, 1966.

Soldiers prepare to fire the 90mm at Bunker 36, Camp Radcliff, 1966.

Guard Tower Number 36 on Camp Radcliff perimeter, 1966.

B Troop 1/9 Cavalry Scouts on the Ho Chi Minh Trail, 1966.

Dixie, a Red Cross Donut Dolly, plays cards at the Red Cross Center in An Khe. There was always a pot of good hot coffee and conversation; a good place to unwind, 1966.

Skytroopers from B Troop 1/9 Cavalry repelling at base camp. All new troops were tested, 1966.

A Troop 1/9 commander arrives right after a firefight near Bon Song, 1966.

Army Republic of Vietnam (ARVN) soldiers advance into the bush on patrol with American troops in Bon Song, 1966.

Pinned down and calling for fire support, UH1 Huey gunship equipped with 2.75 Aerial Rocket Artillery came to the rescue.

1st Cavalry Artillerymen firing on grid with 105mm howitzer at LZ Two Bits, 1966.

Medics tend to the wounds of an enemy soldier at LZ Bird, Dec. 28, 1966.

The fight at LZ Bird, Dec. 27–28, 1966. There were many losses on both sides...a horrible battle that left combat photographer Robert Griffin shaking when it was finally over. This is the aftermath.

Skytroopers clean and repair weapons belong to the dead and wounded at LZ Bird, Dec. 28, 1966.

Soldiers morn the loss of their brothers at LZ Bird, Dec. 28, 1966. The packs are from our dead and wounded.

Villagers at An Khe market in 1967.

Troops were entertained while getting their haircut ir An Khe when the Cavalry Band practiced for music.

D Troop 1/9 Cavalry sweeping for mines near Bon Song, 1967.

B Troop 1/9 Cavalry searching for booby traps in Bon Song, 1967.

Hiking through Bon Song in 1967, C Troop, 1/9 Cavalry.

1st Cavalry Division Artillery softening area so Blues can begin their assault, 1967.

C Troop 1/9 Cavalry check out village in Plei Me, 1967.

Banar tribeswomen and children wait to see Army doctors, 1967.

Infantry soldier working his way across a creek in the A Shau Valley in 1967.

C Troop 1/9 Cavalry Scout preparing to blow up an enemy pillbox, 1967.

Chopping our way into a three canopy jungle and hoping to make it booby trap and ambush free, 1967.

Calling for smoke round prior to firing for effect.

SFC Henley, C Troop 1/9 Cavalry, calling for air support during a firefight.

Troopers burn an abandoned village in Bon Song to find tunnels, and leave nothing behind for the enemy to use.

UH1 Hueys arrive to pick up Blue Team 1/9 Cavalry after a firefight.

Staff Sgt. Eddie Cooper finds VC trail direction marker just outside of Kontum, 1967.

C Troop 1/9 troopers patrol Trail Section II on the Cambodian border.

B Troop 1/9 Cavalry work their way upstream in the Central Highlands.

Walking the Ho Chi Minh Trail in 1968; these were six man patrols.

Banar Tribe refugees try out American cigarettes, 1968.

D Troop 1/9 Cavalry platoon leader calls for artillery support and awaits orders to move out, 1970.

I spoke, "Doc, the pain is coming back, shoot me again okay?"

The doctor replied, "Okay and I'll tell you what, I'll give you another shot. In the morning, my medics will shoot you again and give you two vials of codeine and two needles for you while you make your runs. Don't attempt to inject yourself. Find any medic on the base camps or firebase and they will inject you. I want to see you every morning before you leave and again when you return. We need to check that foot and re-tape on a regular basis, okay?"

I replied, "Doc, you have a deal."

So, for the next ten days it was three hits of codeine every eight hours. The next morning to push the clutch in I used the heel of my foot, turning the foot to the left. The bandage on the left arm was changed every twelve hours by the medics until no bandage was required.

After about seven days, Jenkins returned to Duster Compound after changing a Duster engine on a firebase. He immediately admitted he held a rifle to my head after being questioned by Chief Silvers and Captain Ben. Silvers brought in another captain and the sergeant major.

Silvers called me into his office. I reported and saluted.

The chief, sitting behind his desk, spoke with all others standing in a circle around Jinxie. "Specialist Jenkins has stated he held his battle rifle to your head to force you to drive away from the ambush on the highway eight days ago, is that true Specialist McCormack?"

I spoke, "Aw, chief, you know how he is, just a damned old drunk. He was probably hallucinating, or imagining things."

The chief said, "Is that your official statement, specialist?"

I answered, "Well sir, I reckon so."

"Specialist, stand fast, the rest of you are free to go," the chief stated.

Captain Ben, the last to exit the office, responded to the chief's request, "Captain, please close the door on the way out." The door closed.

"Mack, why are you covering up for that cowardly piece of shit," the chief asked.

"Well sir, he is just scared to death. I am going to have a talk with him man to man. I think I can straighten him out, and who knows maybe he will grow a pair of balls."

"Specialist, you are dismissed; if he ever pulls another stunt like that again, shoot him or I will," the chief said.

That evening after duty we were both were in the hooch, neither having greenline or tower.

Jinxie came down the hall and stuck his head in my door and asked, "Okay if I get a beer from the fridge?" I was cleaning my AR-15 sitting on my cot. I laid down the AR-15 and stood saying, "You chicken shit, if you ever pull another stunt like that again I am going to shoot you myself." Then I paused. "On second thought," calmly pulling the eleven-inch blade strapped to my right arm, "I'll not shoot you but castrate you like back on the farm in Kentucky. Then I'll pour the turpentine to your ass and sew up your bag with the canvas repair kit." Shaking my head, "Then you will never have a woman again, and never have kids; get that Jinxie?"

"Yes," he said and walked away.

About an hour later he came back for the beer saying, "I am sorry for what I did."

I reached into the fridge and handed him a beer, then said, "You think you will go to Long Binh jail for a few months, then back home with a felony and go back to work in daddy's body shop and all will be well, it's not going to work that way. You are going to be a man and fight and kill the enemy, or you are going to get your balls cut out. Then you can become the real meek coward."

Jinxie replied, "Well, if I'm going to jail, I want it to be for something that matters. If anyone messes with you, I mean anyone, then they will have what they need to put me there." Jinxie walked away and from that day on he became a real man and fighting soldier.

CHAPTER FIFTEEN

High Heels

Just a short time later—a normal day—and the old French garbage truck came in on their regular schedule. One thing that was different was that a beautiful female was hidden behind the seat in the cab of the truck. After clearing the gate, she exited the truck as it went around the loop of the berm road to pick up garbage from fifty-five gallon drums spread about 100 yards apart.

She asked questions of the enlisted men going about their duties: "Where does the colonel live? Where do the officers live? Where do the noncommissioned officers live? Where is the entrance to the underground war room? Where is the radar located?

She had long, black hair, wore makeup and was finely dressed down to high heeled shoes. An observant captain grabbed three NCO's weapons, saddled up in his Jeep and chased the old truck down. He caught up and passed the truck. The Jeep blocked the road, and with three battle rifles and the captain with his .45 captured her.

They restrained her hands behind her back and took her straight in to Long Binh to MACV headquarters.

There was an emergency formation in the center of the compound, and the sergeant major asked the questions, "How many of you men told her where the colonel lived? Please take one step forward? How many of you men told her where the officers and NCO's live? Please take one step forward. Half the enlisted men had stepped forward.

"She had this place mapped out in every detail. Some Saigon government people and the CIA are interrogating her. She is a major with Hanoi's finest that are compatible with our CIA," the sergeant major explained.

"Gentlemen, from this day forward you will answer no questions, and if anything or anyone ever manages to get on this base again we must immediately take action! All of our lives depend on that!"

He continued, "Hanoi had a major assault, with sappers planned for us. That is all, dismissed."

Soon after, Chief Silvers called me in early before I would go on commitment and run Thunder Road to firebases and where needed. "Mack, these two boxes are what are called Red Cross sundries pack."

He opened one of the boxes for me to view. Towels, bar soap, razor blades and total hygiene items for combat arms troops in the bush and field, two cartons of cigarettes and chewing tobacco. There was a softback book and a Webster dictionary, one book on getting a High School GED and five books on college freshman GED. You see when a trooper is short (little time left in country), he could go to Long Binh and take the tests.

"You are the only one in my platoon I trust to take these two sundries packs, and give whatever from these packs to the Duster crews in the south corners. They take what they want and the platoon can divide the rest under your supervision."

"Yes sir, chief," I answered.

"Oh, as far as the platoon goes, you have first pick on what's left."

"Thank you, chief," I said.

I carried the packs down to the first Duster and said to the track commander, "Hey, you guys can take whatever you want from the packs."

The commander looked through the packs and removed the carton of Salem cigarettes and said smiling, "This is good trading material in any village; these Salems' are worth ten dollars and the prostitutes will trade for them. He removed the toothbrushes, toothpaste, razor blades and like hygiene items. He picked up a romance novel and said, "We will take turns reading this one, we love these novels."

"Anything else you want?" I asked.

"Not really," the track commander commented.

"Good, I want these GED books, my high school education stinks, and I've already made the decision that if I manage to live through this, I'm going to get all the education I can, so if a world war comes and I'm recalled, I can be in the States and never have to do any more of the shit like we are going through, enough is enough," a crewmember stated.

"I've already decided I will do whatever it takes to teach auto mechanics in a technical school back in the World," I told the commander.

We parted ways, and I went to two other Dusters and did a repeat with the second gunnery crew.

The next morning Chief Silvers called me into his office. "I hate to do this to you, but it's your turn and time to attend the ceremony on the Parade Field for the fallen. You will be present for the awards presented to the living."

"Just stand at attention; tune out what is being said. If you pay attention it can mess your mind up. You have to be numb, as I have told you before you just can't have any feelings as to what really goes on here," chief added.

We assembled and fell in formation for the ceremony. There was a row of bayonets buried into the earth. For each fallen brother, there was an M16 attached pointing down, of course, with a steel helmet with a camouflage cover on it. The name of the fallen trooper was very neatly stenciled on in black letters.

The lieutenant colonel began to read the names, rank and such for each of award recipients. He announced the Purple Hearts and awards for uncommon valor. Then there was the 21-gun salute with live ammunition from M16 rifles.

When that portion of the ceremony was complete, two searchlight men were decorated, each receiving the Purple Heart, one receiving the Bronze Star, and the other the Silver Star for uncommon valor.

The lieutenant colonel had brought in a Quad 50 crew of newbies who had been in country for less than 30 days for a reason.

When the ceremony had ended and we were dismissed, the searchlight team members who were awarded the Silver Star approached the newbie Quad crew and said rather loudly, "You people are cowards, pure chicken shit." He finished with, "You people keep away from us!"

The sergeant of the Quad crew started to speak, and was interrupted by the searchlight crewmember.

"Shut up, no excuse can change what the four of you did!" The two of them walked away.

I followed them to the little building close to the east berm that belonged to the I-29th searchlights. I asked the two searchlight team members what that was all about, and they went on to explain, "We were in the tower and the VC hit us hard, about a battalion of them. They managed to get in close, that crew couldn't hit anything with their guns. The crew cut and ran for the command bunker with just their M16's."

"The VC were all over our tower trying to lay satchel chargers to blow the posts out from under us. We had our sixty and some cases of grenades. One thing we do, because there is normally only two of us in a tower, is to drive some nails in a little and bend the end of the nail up ninety degrees so you can pull a grenade pin with one hand while feeding the 60 with the other.

"We both were shot through one arm, I was firing with only one good arm, and he was alternating with his one good arm, feeding me and pulling grenade pins with the one good hand.

"When those cowards reached the command bunker, the captain put his .45 in the sergeant's face and told him, retake that gun truck, mount it, fire those guns or I'm going to blow the brains out of all of you here and now. Well, they did retake the gun truck but we weren't evacuated until first light. A medic made it into the tower, and we held out until first light. We just got out of the hospital you see," said the searchlight crewmember.

CHAPTER SIXTEEN

MACV

The lieutenant colonel called Chief Silvers into his office and gave him pure hell for holding retrogrades too long and not getting them soon enough to Guam for rebuild. Chief Silvers wasn't going to take that. Early the next morning he called me into his office and requested that I close the door behind him.

Chief said, "The colonel is all over my ass for the way I'm running things. I'm having some components flown today so you can take me to MACV. There are some things I want to show you; if you do not see them you and the people back home would never believe it. Get your weapon and gear and be back in five minutes."

"Sir, yes sir," I replied.

Chief and I cleared the gate. He had his khaki dress uniform and slippers laying in the rear of the Jeep. "I am having you to take me to MACV to request a transfer to the Advisory Force. I did that back in '65. I was also an advisor in the Korean War my first tour as an enlisted man."

"Say nothing to anyone; I know you will not," chief said. "The colonel doesn't know much of how to keep those machines (Dusters) running or firing. For example, if we order breechblocks through channels it could take weeks to receive them, and in the meantime a lot of young men will die."

We entered Long Binh. Chief Silvers had me stop at a unit close to the MACV so he could go shower and change into his khaki dress.

I thought, "My God, awards from the pocket to the shoulder tab, airborne, heliborne, and the combat infantry badge, multiples, so many unit citations." I took Chief Silvers on through the security gate there at Military Assistance Command Vietnam. Chief stepped out of the Jeep and said, "You stay here with the vehicle; do not let this machine out of your sight as I have taught you. One has to be in dress uniform to enter MACV. I won't be long."

In no more than twenty minutes the chief returned, climbed back into the Jeep and said, "I need to go back and change, then we will tour some places here in Long Binh."

After changing, the chief directed me to the northeast entrance to the berm road of Long Binh. "I'm going to inspect the bunkers and berm of this base camp; it will be something to tell about back home someday."

It was 33 miles around that huge base camp. We stopped at every bunker, and the chief entered each and every bunker for one or two minutes. Eventually we had made the complete circle.

The chief looked at his watch and said, "Let us go to Binh Hoi for a noon meal, my treat for the top young man in my platoon."

After entering Binh Hoi the chief instructed, "Pull in there, the officer's club."

As we entered the club, a SP-4 with an apron around his waist said, "Hey chief, you can't bring an enlisted man in here. This is for officers only."

"Oh yeah, I want to see the officer in charge!" Silvers screamed.

A major walked up and said, "The enlisted man can eat at the enlisted mess."

"Look major, wherever I go this trooper goes with me!" Silvers screamed.

"Not going to happen, chief, this is officers only," the major replied.

Silvers screamed, "I want to see your bird here and now! This man runs the bush every day! I think I can convince your bird and command sergeant major to order you to ride with him shotgun and work the 79, lob a few grenades? You special services people need to learn what war is really like! I've got people dying out there! I want your bird in front of me now! I'll have a talk with the man that has stars on his lapel about you, base camp warrior!"

The major turned ghostly white in his face, "Okay chief, he can stay."

"From now on when he walks in here, feed him!" the chief screamed.

The major replied, "Okay, for you, okay."

The chief sat down at a table and addressed the SP-4, "Bring two T-bones, two baked potatoes, two salads, and by the way, maybe I'll talk to the bird that you need some combat experience and can ride shotgun with this man."

The SP-4 only glared at Chief Silvers. Then asked, "How do you gentleman want those steaks?"

"Make mine medium rare. How do you want yours little man," the chief asked.

I replied, "Medium well, no bleeding. I only want to see Charlie bleed, and lots of it."

The chief began to talk and I listened. "You know your scores on entering this man's army? You no doubt scored in the top ten percent of mechanics, I'm sure. You did tops in some things that could have made you a fine officer."

"Well chief, it was offered three times at Polk, but you know I only want to be responsible for me. I could not live with myself if a decision I made cost someone their life," I confessed.

"Yes, I have some things in the past. As an NCO and officer, it haunts me some of the decisions I was forced to make." Then the chief looked away with a very strained look, and became very quiet for a short time. "Some of the men in the platoon have said you want to return to the States, re-enlist and become what I am, is that right," the chief asked.

"Thought about it some, chief," I replied.

"We've already talked about you becoming an auto instructor at the school in Elizabethtown. I'm putting you on notice, re-enlist and I will find you, beat your ass so bad you will walk sideways for a month. Hell, you will just be sent back over here, and you will wind up like me, shell shocked and messed up in the head," the chief admitted.

"You are getting out when your ETS date comes up, and I am ordering you here and now to do whatever it takes to become an instructor." Chief continued, "I know you will do fine if you live through this insane war. You need to learn how to make a living and to rejoin society. Society has no use for trained killers. That is when you come in, that will be your new job," the chief commanded. "Well, here comes our steaks. Enjoy, there is something else I will show you today," the chief said before becoming very quiet.

After the meal, we went back to Long Binh with the chief giving directions.

"You are going to see what no enlisted man has ever been allowed to see." We stopped at a checkpoint and had to present our military ID cards. "Strange — never had to do this before," I thought.

Another mile or so coming over a slight hill, I was in awe.

Silvers spoke up. "This brick house here in the center of Long Binh is known as the house Westmoreland built. The new theatre; commander has his wife and teenage children live in that house.

"Look around, the area surrounding the house is an 18-hole golf course. You have to be a bird or better to play golf with the general on Sundays," the chief laughed.

"Chief, that thing next to the house looks like a silo on the farms back home, what is it," I asked.

"That is a nuke shelter for his family. The cook and the maid and all in the house if Peking or Moscow decides to nuke the place," the chief answered.

"Turn around at that MP checkpoint. He would never allow us through that last checkpoint, because I don't have enough eagle shit on my collar, anyways," the chief commented. "Head out and I will direct you to the last thing on Long Binh I want you to see today."

Giving directions, we came upon what was some kind of racetrack. "This is where each unit here on the base camp races their shop built go-carts on Sundays. They steal small engines from us and other units when we turn them in for repairs. We just never get them back. That is what happens to some of our 'Little Joes' for the Dusters and Quads," the chief admitted.

"The base camp brass always brings the Donut Dollies and whatever round eyed nurses to impress them," chief commented. "Let's go by II Field Base, then it's back to Duster."

"Take us to that bunker there, there by the gate, and MP checkpoint," the chief pointed. Nothing more was said as the chief exited the Jeep and walked into the bunker. He just stared out the front in the general direction of Binh Hoi. Then he, for about a minute, stared down at the floor. "Come here into this bunker." He paused and then said, "This is where I spent 72 hours during the siege in TET '68. You probably saw it on the six o'clock news," he continued. "The Viet Cong was in between us and the troops defending Binh Hoi. Our soldiers were killing each other from fire trying to kill the Viet Cong. A major, a first lieutenant and myself defended this position. We rarely had time to eat, with no sleep during that 72 hours. We destroyed three barrels for the 60, no time for barrel change. Sometimes the 60 would cook off, but we'd keep firing because the chamber would be so hot when one releases the trigger she continues to fire. You stop it by twisting the belt. There were charred burn marks on the wooden floor from empty brass that were so hot they burned into the wood."

The chief pointed out of the bunker to an asphalt patch on the street, "See that? An ammunition truck that just drove by throwing out ammunition and grenades took a shaped heat charge from an RPG. All that was left was the truck frame and rear axle housings. It blew out the engine, transmission and even the transfer case." After a long pause he continued, "They never found anything of the ammo truck or its crewmembers!"

The chief took two steps to the front edge of the bunker, stretched and reached out touching the ground in front of the bunker, "I killed my own barber. He had cut my hair on both tours and had security clearance to come onto II Field Base."

The chief shook his head in disbelief. "We piled up the dead VC so deep you could hardly fire rounds for the dead bodies. At times, we had to shoot through the dead to kill the living, who were doing their best to kill us."

"Some day you will have to tell the World the truth of what happened here. They were coming at us running over the dead bodies," chief said with his head turned, looking at the dead back in '68.

We returned to Duster Compound. That night was just a typical normal night on Duster. At dusk the greenline was manned with three-man fire teams, plus the normal crew of three men in the watchtower. The Dusters were working out. One-gun tube on one Duster firing semi auto, fire two rounds, wait three seconds, and fire three rounds, switching off gun tubes. Then after an hour the other crew would take over.

Charlie, in two and three-man recon teams, would probe the perimeter for weakness. The 3/4 Cavalry of the 11th Armored Cavalry had already dropped in their three-man ambush teams and already set up the auto ambush with claymores and high explosives. Yes, just a normal night on Duster Compound.

The war room would call a bunker and inform the fire team, "Charlie, three of them at 10 o'clock 200 yards, fire one grenade for the effect or two for kill."

But when I was on the greenline, I would fire 12 grenades.

The chief walked up to my platoon's hooch.

"Come outside here Mack," the chief ordered.

Jinxie had finished most of a fifth of cheap bourbon and was passed out and sleeping on the ground.

"Look at all of this shit. He's passed out drunk. If this place is overrun, Charlie will bayonet him, he will die and never know what happened. There are about two dozen enlisted men in that so-called club drinking and Charlie is probing the greenline!" Shaking his head, "There are men in the little theatre watching an old movie eating popcorn. This place is going to be overrun and wiped out." His voice got louder, "There is a breakdown in command here. Someone will have to tell them what went on here, someone has to live to tell that."

"Go down to my bunker, get an empty 100 round ammo can for the 7.62 x 51MM machine gun rounds and fill it half full with grenades.

Then fill it up with 20 round magazines for your battle rifle," the chief ordered. "There in the corner of the blast wall, pull out a couple of sand bags and hide the can there." He added, "When Charlie penetrates this place, just grab your rifle, bail over the open side wall of the hooch and hunker down in that corner. Don't fire on them unless they spot you. Then try to get over to the Special Forces side."

"Someone has to live to tell what went on here," the chief said. "You've got to live to become the instructor at the school!" Then the chief turned and walked away.

Two days later Chief Silvers' orders for his transfer were approved. He loaded his duffle bag and gear in the Jeep. He said his goodbyes to Captain Ben and the men in the platoon. "Little Mack, go and get your weapon and gear to transport me to Long Binh. Please bring your Colt rifle for me to use. Sure hate to be caught unarmed on the last trip in. I had to turn in all my weapons."

"Sir, yes sir, chief," I replied.

The chief never spoke a word going to MACV. He exited the Jeep sat down his duffle bag and gear. He offered his hand, and the chief and I shook hands. All of a sudden, the chief put his arms around me hugging me and he began to shed tears saying, "You've got to get back and teach at that school. The men are going to need a man like you that will understand them and understand what they have been through."

"Now I'm giving you my last order, I am giving you five years to be there at that school."

The chief turned and walked away, duffle bag in hand.

And indeed, five years later in uniform he would walk into that school to check and be sure that I had become an instructor.

When I returned to Duster Compound, parked the Jeep and walked into Russ's office, I spoke to Russ, "You know, the chief broke down and cried."

"Yes, he did the same thing that day you and Jinxie were out on Highway 1, you know when we thought you both were missing in action," Russ recalled.

"I know that whatever the chief endured in '68 really bothered him. I think he felt responsible for that," I revealed.

No more was said.

CHAPTER SEVENTEEN

Crown Royal

And when he gets to heaven,
To Saint Peter he will tell;
One more Soldier reporting, sir.
I've served my time in Hell!

A Duster crew had come into Duster Compound from a firebase. The crew was in bad need to go to the II Field PX and other errands on the base camp. I was asked to pull watch in the gun tube, as I had been trained by Chief Silvers. In fact, the whole platoon had qualified the 17B20 military occupational skill. I loved the duty as well as using high explosives, the grenades and claymores.

One crewmember remained with me in the gunner's seat; he had just taken a shower and had only a towel wrapped around him, flip flops and his boonie hat on. He was sitting on an ammo can on the ground eating a tray of food.

"This chow tastes like shit," he said. "Well, as hungry as I am, it's not so bad for shit."

I was sitting in the gunner's seat carrying on a conversation about the World back home. The field phone rang, the man in the tower screamed, "Gooks in the wire, gooks in the wire!"

I had been looking to my right. I screamed, "Gooks in the wire, gooks in the wire!"

The gunnery crewmember jumped to his feet frantically, threw the tray into the air, chow going everywhere. Wildly he went up the side of the Duster, lunged into the gunnery seat as I jumped to the right to load. The twin 40mm gun tubes roared at the rate of 240 rounds per minute. I glanced up at 12 Viet Cong charging at us with the AK-47 fixed bayonets, Chinese vest with the magazines, each having the grenades tucked under a belt with satchel charges strapped on their backs.

Duster rounds firing at the enemy only 100 feet from the Duster, bodies of the Viet Cong instantly vaporizing, body parts being scattered as rounds set off the satchel charges strapped to their backs.

The left gun fired dry for lack of a loader. I continued to place the striper clips into the loading tray. The gunner continued to fire full auto. More than 1,000 rounds were fired, raking the gun tubes left to right back to left.

When the gunner ceased fire and the dust was settling I commented, "No bodies, no body parts."

"Hey, our crew vaporizes the commie sons of bitches, nothing left to have to bury," the gunner commented.

"No confirmed kills," I commented.

"We don't give a shit about body count, we just worry about staying alive man, and sometimes that is not very easy," said the gunner.

Two days later I was running firebases and had stopped at one of my spies to hand over a roll of currency. I collected the page of information, folded it two inches square and placed it down into my jungle boot. Duster Compound was flying in components to the base camp Cu Chi and I would take it from there.

After a hard day, I reported to the clerk managing the deliveries flown in. He gave me the components to deliver. When I commented, "Hey man, it's getting late. I need to be back on Duster Compound soon, I have greenline tonight."

The clerk commented, "Here is a bag of trigger bars for the Quad 50 needed bad on a firebase 45 miles south down River Road. You've got to get these trigger bars now or a lot of people will die tonight or a rifle company and the Quad crew will be wiped out to the last man."

I barked, "What in the hell are you talking about?"

The clerk answered, "The Quad is down to one working gun, as three guns have broken trigger bars and no spares."

I roared, "Get a bird to fly them out."

"No one will fly, there is a battalion of VC and they have heavy Soviet machine guns and shoulder rockets that will bring down every helicopter before they can reach the firebase. It would be suicide for a crew to even try. You are their only hope to survive the night," the clerk admitted.

"Hey, put a crew together in a deuce and a half with some 60s and go in!" I urged.

"There is no one on this base camp that will go; you are the only hope for those men not being totally wiped out. You are the only man that has the balls to do a mission like this," the clerk acknowledged.

The clerk held up the Crown Royal purple bag with a dozen or so trigger bars in front of me.

The clerk pleaded, "Please do it, you are their last and only hope."

I fired up the Jeep, and shot into the battery mortar service and screamed at the staff sergeant with the battery commander at my side. "Hey staff, fill my machine with fuel from the tanker; I have to run south 45 miles.

The sergeant hustled with the fuel hose and topped off my fuel tank without speaking a word.

I fired the engine and dumped the clutch, the Jeep spinning in the sand on all four tires.

The captain commented to his motor sergeant and faintly heard by myself, "That kid is an absolute total, f'n lunatic."

The monsoon was about over. I had driving goggles to protect my eyes, and a handkerchief over my nose and mouth tied off with a string. The dust from a deuce and a half was horrible that I had to follow for about 15 miles before it finally turned off.

Now there was a clear sand road ahead. I came upon and passed the old abandoned grain mill with the water wheel lying upon the riverbank. Miles of the road ran parallel with the river. About a mile south of the old mill, two Navy patrol boats were giving everything they had fighting a rather large number of Viet Cong. The gunfire was so intense it was just a solid roar and scream.

One 7.62 round went about 18 inches behind my head. The sound of that round was like the sound of a bumble bee on steroids and supercharged.

Another tumbler came through the Jeep behind me — "rope, rope, rope," as it is called a "tumbler" in slang terms. I mumbled to myself, "It's impossible to get away from this shit!"

As I came up about a half mile from the firebase, it was under fire from the Viet Cong. I picked up the microphone on my radio set and yelled, "This is Duster Chief, I say Duster Chief, I'm coming in, pull the wire."

The response was, "We are hot, we are hot, go back, go back!"

"Damn it, this is Billy the Kid, I say again, it's Billy the Kid, pull the damn wire or I'm coming over it to get in!" The Jeep roared in, making a turn and spraying sand and coming to a quick stop.

There was incoming and outgoing gunfire on the north side of the base. I pulled my AR-15 from the dash mount with my left hand, and held the Crown Royal bag in the right. I started to run to the greenline, when a rather tall and well-built captain grabbed me by the rear shirt collar.

"Son, what in the hell are you doing way out here," the captain asked.

"Trigger bars for the Quad 50, heard you were in dire need," I answered.

Just at that instant the sergeant gunnery crew commander jumped from the truck bed of the Quad 50-gun truck to me. "Thanks kid, you have saved our bacon." The sergeant ran, jumped back on the gun truck issuing trigger bars from the purple bag to his two loaders to install, then the sergeant screamed, "Trigger bars, three cheers for Billy the Kid!"

The gunfire had died down, all the 25th Infantry men jumped to their feet throwing boonie hats in the air screaming, "Three cheers for Billy the Kid, three cheers for Billy the Kid, three cheers for Billy the Kid!"

"Kid, you need to stay the night with me, it's too late for you to be on the road; I'll give you a bodyguard and keep you close to me and the first sergeant," the captain said.

Now Chief Silvers had taught me so well that only the Recovery and Deployment officer had power over me.

I answered, "Sir, with all due respect to what you have on your collar, just screw you, sir." The captain's lower jaw dropped and looked wide eyed at me, then laughed and clapped his hands together.

The captain then said, "You know I could arrest you and put you in irons."

"Captain, then you would have to put two bodyguards on me. Those two need to be on the greenline killing gooks. My machine could very well be shot to pieces. It would take six months, if then even, to get a replacement. Where would that put these gunnery crews. I would just have to go to Saigon and steal one from some dumb jarhead officer. All that work doing a repaint and changing the data plate would cost me time that I need to be running Thunder Road." I looked at the captain, "Sir, I do my best work out there on that road."

The captain shook my hand, pulled me in and gave me a hug. "Kid, keep the rubber down, the steel up. God go with you, and a guardian angel riding in the back seat of your machine."

"He always does, he's the only one that will ride with me," I said as I pulled out for the 45-mile run back to Cu Chi.

I pulled up to the south gate and the MP, who some months before had befriended me said, "Man, we have a major assault coming here tonight, thought I should tell you."

"Thanks, I need to know that," I replied.

I reported in to the clerk at battery headquarters. The clerk said, "Man it's just too late for your trip back to Duster Compound. We will give you a cot in the center of this base and you can get some sleep."

In a strange odd way, I announced, "Cold day in Hell, fella, when I spend the night on this base that has major offensive coming tonight! I'm headed home to Duster Compound."

On the way out, I added, "Hey desk jockey, get Duster on the radio and tell them Billy the Kid is on the way home."

"Okay, but you are making a mistake leaving this late, the roads shut down at 1700 hours and you will be all alone and no one to call for help if anything goes wrong," the clerk said.

"I can take care of myself, I work alone. Army Rangers lead the way; hell boy, I am a Tiger Ranger trained by a man that has won his green beret. Don't get any better or finer than that, Tigers are the best in the world," I declared.

"Hope to see you next time," the clerk said, shaking his head.

"Right on," I barked as I went out the door to saddle up for the trip back to Duster.

I approached the south gate, and my friend, the private first class on the gate, opened the pole gate for me, and waved me on. As I went through the gate, the MP kind of saluted and screamed, "Give old Charlie hell kid!"

As the sun was dropping from the sky, I was running 55 to 65 miles per hour coming up on Phu Loi base camp. Picking up the microphone to the radio and said, "Phu Loi, Phu Loi, this is Duster Chief, I say again, Duster Chief. Hold your fire, don't shoot at me, I'm coming through!"

The only response from the radio operator in Phu Loi was, "Roll on Billy the Kid, roll on!"

I was doing good until about 15 miles out of Binh Hoi, when the Jeep began to vibrate violently coming from the right front. I said a silent prayer, "God help me, this thing is losing a front axle, just help me get through this."

I slowed down, coasting, and flipped the passenger seat forward reaching into the well under the seat into my tool bag with my right hand, coming out with pliers and a small coil of lacing wire. I was coasting down and shut off the master switch for total silence and stealth.

I steered the Jeep over the deep ditch line and straddled it. Grabbing the AR-15 and one M56 from the bag, I ran around the rear of the Jeep and crawled under, observing that the outer CV joint had come apart with the axle just hanging down.

With the AR-15 lying across my chest, the grenade lying in my crotch and my combat knife strapped to my right arm, I quickly tied the loose axle up and out of the way. As I was finishing the tie, I could hear movement about 10 yards out in the elephant grass and a silent prayer went out to my beloved God and guardian angel, "God, please let what I hear be an animal."

I crawled out from under the Jeep, tossed the AR-15 into the passenger side, stepping quickly to go around the rear of the Jeep. As I was at the rear of the Jeep an old Viet Cong, about 60 years of age, stepped out of the elephant grass about six feet from me, AK-47 in hand. Expecting the worst and death, I screamed out to the old man as he was turning the AK-47 on me, "Hey papa san, we die together tonight, you for Hanoi, me for, the great and powerful U.S.A. We will open the gates of Hell together!

I had pulled the pin on the M56 grenade holding it in my left hand. My hands up as one would surrender; however, I knew I could not be taken alive. No one human could take the brutality and torture the Viet Cong would do to any man.

"I'm going to put this grenade down your trousers and blow your balls off, old man!" I screamed.

The old Viet Cong stood frozen, face ghostly white, frozen for the moment in time.

After about three seconds, I slid around to the driver's side of the Jeep. I literally dove in, flipped the master switch on, fired up the engine, revved up the little 4-cylinder engine and dumped the clutch, and launched the Jeep, spinning the rear tires on the wet elephant grass.

The grenade was still in my left hand being held with two fingers and thumb, the pin still on the center finger — you know, the one that gives the southern California greeting. I ran out in low gear and went to second gear. At about 35 miles per hour I made an attempt to move back onto the road. The Jeep skidded, so I recovered and stayed over the ditch line. Another attempt had the same results. The third attempt, with more speed, was a success. The Jeep launched, with tires almost clearing the road. The Jeep swerved three times, and I corrected.

Grenade still in hand, I questioned myself, "Should I just throw this grenade or try to reinstall the pin?

Well, on a long straight stretch of the road, with both knees under the steering wheel to lock the steering wheel straight, I quickly reinstalled the pin.

About a half mile from Binh Hoi, I picked up the microphone to the radio and said, "Binh Hoi, Binh Hoi—this is Duster Chief, I say again Duster Chief. Hold your fire, check your fire it's me, I'm coming in."

There was no response.

As I pulled up to the barriers at about 2100 hours, the two 113s and the Quad 50 from Delta 71st pulled the slides on all M2 .50 machine guns and a squad leader screamed, "Halt! I want to see both hands out the door or we will blow your ass away! I want to see an ID card NOW!" the squad leader roared.

I took out my ID card, and put both hands out the door opening as far as possible, I was challenged for the password. "I was not given the password on firebase or Cu Chi!"

"Who won the World Series past play off," asked a voice from the darkness.

"I don't know who won, I don't give a damn who won, I'm not into baseball all that much, I'm into other things," I offered.

"I'm coming to you to take a look at your ID card, make one wrong move and these meat grinders will make sausage out of your ass," commanded the sergeant.

As the sergeant approached the Jeep, he turned on the GI flashlight in his left hand, a .45 in his right pointing at me.

The flashlight beam in my eyes I roared, "Get that light out of my eyes. You are killing my night vision fool!"

The sergeant commented, "What are you doing out after dark, man?"

I explained about the dire need of trigger bars on the firebase south of Cu Chi. Then a loader jumped from behind his bat wing and cried out, "Hey I know this man, he is from Duster Compound. This man is Billy the Kid!"

The sergeant handed me my ID card back and said, "Hey, we're so sorry, we thought you were a Russian Spetsnaz son of a bitch impersonating one of us and up to no good. Hey, pull the barrier and let him in to go home. I'll call the next road barrier and they will wave you on through."

It was approximately 2130 hours when I called Duster Compound to make the officer of the day aware of my very late arrival back home.

Driving into the motor area, I immediately put my grenades in the platoon's red status bunker. I was walking back to the Jeep and it hit me; grabbing onto the side of a retrograde, I threw up everything inside of my stomach and then some.

The officer of the day walked up asking, "What happened out there, trooper?"

"Nothing sir, must have been spoiled food at the noon meal, sir," I replied.

"Bullshit! I know better than that, some garbled reports of you on the radio. You were supposed to have greenline but I put one man from your platoon on. Have a couple of shots of that vodka you keep, relax, and calm down. Get a good night sleep, you have another full day come morning."

"Yes sir, I really appreciate what you are doing for me sir," I answered.

"Thank me and Chief Silvers by getting your ass out of this insanity and becoming that auto teacher that Chief Silvers was and is so obsessed that you become later in life," the officer added.

"Sir, yes sir," I replied.

"Well, that is all you will ever be asked to do trooper, now go and get some sleep."

"Yes sir, thank you sir."

I continued to go in and out of firebases, and many times I arrived just before or just after firefights. On occasion I visited the four good and dependable spies. Soon I was told not to stop at the location where the young lady washed clothes but was not told why, just, under no circumstances, to never stop to attempt to see her again.

CHAPTER EIGHTEEN

Chief Baltier

It was now official, the scuttlebutt rumors were true — the 5/2 would be standing down and leaving Duster Compound and turning the entire compound over to the 30th ARVN Rangers. It was announced that in mid-1971, the entire battalion would stand down and return to its stateside home at Fort Bliss, Texas, the home base for air defense systems. The base would still be held for some time by Head and Head by trucking what would be the "Day Crew" and the "Night Crew" to the base.

The crews of the Dusters had already begun training crews from the Saigon government on maintenance and firing the Dusters.

The recovery and deployment platoon had gone some time without a warrant officer, a captain, or the sergeant first class NCO. The first to arrive on Duster Compound was Chief Warrant Officer 3 Baltier. He was Chief Silvers' replacement. He was short, chubby, and had an olive complexion of west European decent.

Only two days after his arrival, I transported him to the line battery at Cu Chi. The trip up was uneventful. He spent some time with the battery commander and first sergeant. On the return trip to Phu Loi, I was shocked as the new chief placed his left hand in my crotch.

I pushed his hand away and responded, "I wouldn't do that if I was you, chief."

A couple of minutes later the act was repeated, "I'm not going to tell you again chief, I'm not like that," I said as I pushed his hand away the second time.

The chief said, "I can see you get that Purple Heart and that Bronze Star for valor, maybe I can have it upgraded to the Silver Star; I can make life so good for you."

"No thanks, chief, not even for the Congressional," I blurted.

The hand returned for the third time.

I turned and threw a karate punch to his nose as the chief was looking me straight in the eye. The chief blurted out, "You little son of a bitch, you have broken my nose!"

"Hey, I don't put up with that kind of shit!" I roared.

"I'm going to the lieutenant colonel on you for this; you are going to do some years in Leavenworth for striking an officer!" Baltier screamed.

Down the road, the front drive unit started to howl. At the intersection at Binh Hoi I was forced to return to two-wheel drive to prevent total failure of the front drive unit. We entered Duster Compound and the chief screamed, "Stop, I'm going to the battalion commander on your little ass; you are going to Leavenworth for some years!"

Well I figured, "If I'm going, I might as well go for something worthwhile."

The chief made a big mistake of trying to walk in front of the Jeep. I floored the accelerator and dumped the clutch, just as Chief Baltier tried to cross in front of the Jeep. Bam! — a shattered knee cap and broken bones in his foot, as the right front tire ran over his foot.

The chief screamed, "You little son of a bitch; you ran over me!"

"Payback's a bitch!" I roared.

The chief struggled in a half crawl to get to the command post. I rushed to motor service and parked the Jeep. I hustled to Sergeant Timolsic, the first person I could find, and reported the event. Sergeant Timolsic hustled to the lieutenant colonel's office and told the story.

The lieutenant colonel said, "I won't do anything to him this time. I've had previous reports of similar happenings from other troops. If McCormack touches him again he will be arrested and court-martialed."

Later that week, the platoon of enlisted men was taking their ten-minute break. Each man was very solemn, when Jinxie asked, "Any reason you guys might think the chief's a queer?"

The whole platoon busted out in a roar of laughter and each told of their incident with the chief. Jinxie admitted, "Well, I just caught him in a hooch going down on one of the captains."

There was hilarious laughter. After that, the captain in question was harassed by both the enlisted and some of the NCO's.

Two days later, I caught duty as driver for the officer of the day, 24-hour duty. This was my first time as I had been exempt when Silvers was on base. I believed it was punishment for what I did to Chief Baltier.

Chief Baltier stopped all awards including the Army Commendation Medal put in by Chief Silvers. The orders came through and were stopped. He even tried to take away my Good Conduct Medal; however, I had already received the orders and did retain that award.

Early in the morning about 0200 hours or so a fire mission was to be cancelled on the southwest Duster. The captain said, "Come on kid, let's drive the 3/4 and you tell the crew to cancel."

We drove the short distance in the 3/4 truck. The captain said, "Run up the sandbag steps and tell them to cancel the fire mission."

I did so, and as my face was about 18 inches from the muzzles of the twin 40s, the sergeant fired the guns. The muzzle blast literally blew me over the side of the blast wall, burning my face, burying cordite (gunpowder used in the artillery gun) in my face and burning the retina of my left eye completely away.

I was rushed for treatment to Dr. Soltis and the medics. I was not relieved from duty, but merely wore an eye patch for ten days.

The retina re-grew; however, it took on a different shape and stigmatism than the right eye. My vision was lost in that eye then and there at that incident.

CHAPTER NINETEEN

"One, Two, Three, Four, We Don't Want No F'n War!"

Just before the Thanksgiving holiday, a new captain arrived to replace Captain Ben. It turned out this man had been deferred by the draft system so he could first attend a university for an associate degree to become an officer. He had failed to make an acceptable letter grade of a "C" in his English 101 and had to retake the course, but did pass and receive the "C" by only one point.

In time, all would learn the man did not have the ability to command. When asked a question by an enlisted man he would say, "I have to give this decision some thought and get back with you."

He would then run to Chief Baltier or some other officer and ask for advice on how to handle the situation; literally, what an incapable ditz!

During the move, myself, another enlisted man, and Sergeant Timolsic had our personal gear, our steel folding cots and such on a deuce and a half, plus some "throw away" items. We pulled into the laundry pit, or in civilian terms — the landfill — and were off loading throw away items when a 16-year-old cowboy jumped onto the truck attempting to push me over the side.

I grabbed the main frame of a cot. The cowboy slammed the folding legs closed, shattering three fingers on my left hand. I screamed out from the pain. The cowboy jumped from the truck when I reached and grabbed the AR-15 and attempted to shoot him right handed (I am left-handed).

At that instant, a Huey was hovering overhead. The Sergeant Timolsic pushed the barrel of the AR-15 into the air and screamed, "You have eyes on you, do this and it will be Leavenworth! You will have another opportunity later."

After three days of extreme pain, the sergeant took me to Long Binh Hospital. Some female locals were being trained under American nurses and surgeons to replace the American nurses.

Sergeant Timolsic accompanied me into the emergency room after the X-Ray. The surgeon asked me, "What happened, someone make a mistake with a sledge hammer?"

"Something like that, Doc; it's a long story we just do not have time for," I answered.

Sergeant Timolsic cut in and disallowed me from saying anything else.

In resetting the fingers, I went unconscious. When I woke up, I panicked while a local nurse at my side. With my right hand, I karate punched her in the center of the chest, knocking her down in the sitting position against a wall. The surgeon screamed, "I could have you court-martialed for doing that to her!"

The sergeant cut in again, "We have just come in from the bush. A cowboy broke his hand three days ago, and he has been in a lot of pain for that time; we are so sorry for what just happened."

"I want to hear it from him!" the surgeon demanded.

The nurse standing against the wall, as far back as possible had a look of absolute panic and fear on her face.

I looked straight at her face, "I'm so sorry, ma'am. It's been a rough and painful three days."

"Okay, I'll let it go this time. If you ever assault one of my ladies again I'll put you in Leavenworth."

I replied, "It will never happen again, Doc."

"Better not," the surgeon grumbled.

It was 2345 hours and remember, I was courier for the 5/2. At the main gate, the CIA and their two 113s were present. The lieutenant colonel, the sergeant major and all the captains were present. A captain connected to the CIA requested, "Colonel sir, give us your best courier for a mission."

"That would-be Specialist McCormack, captain," the colonel replied.

"Specialist McCormack, step forward please," commanded of the CIA officer.

"We have a mission for you in ten minutes; all you need is your Jeep. We have a .45 for you and it's all you will need. Report back here at 1200 hours, ten minutes from now in your Jeep," ordered the CIA officer.

"Sir, yes sir!" I responded.

"Damned fools think I will go down any road this hour without my AR-15 and grenades they are bigger fools than Santa Claus could ever imagine," were the thoughts in my mind.

As I rounded one of the buildings, one of the young men that worked in the radar/radio caught me by the arm, crying hysterically.

"What the hell is so bad fella?" I asked.

"They have set you up to be killed so they can capture a VC colonel. They are sending you on the military road to II Field Headquarters with a briefcase cuffed to your arm with nothing inside but blank pages. About two miles from the 199th Light Brigade, as you slow down, the VC will ambush and kill you. Then MACV Rangers will swarm in and capture them, but you will be dead," the young man unloaded.

"Dry up, pretty boy, not going to happen tonight. Hide and just watch what happens tonight, these dumb shit officers will not get me killed," I replied. "I never have gone all the way on the military road at night anyway. I am going to love the look on the faces of those shit faced sons of bitches when I get back," I declared. "I've got to hustle, thanks for letting me know. I'll never forget you for tipping me off."

I doubled back to the hooch, grabbed the AR-15, the Duster ammo bag and 20 rounds for each magazine. I hustled to the greenline bunker and grabbed six M56's in a bag. I placed the AR-15 ammo and M56's beneath sand bags under the back floorboard.

I quickly unhooked every light on the machine except leaving the right front headlight functional. I saddled up, fastened my custom seat belt, and roared around the berm road, throwing sand. The Jeep skidded a little.

"Here son, a .45 and two extra magazines; you shouldn't have any trouble," said the CIA officer.

They cuffed an attaché case to my right wrist. "Pull the gate," I shouted.

As I roared out, full throttle, one of the officers hollered, "Turn those lights on!"

Out the left side of the Jeep, the left arm and hand as high as possible with that southern California greeting of the single middle finger.

"Nice moon and starlight tonight," I thought.

As I approached the old French fort, I turned on my one single headlight.

There was a cut road from the military road to the local back road that came into the rear of Duster Compound; the entrance where I was brought in on my first day. I pulled up to the roadblock with my own D71st Quad 50 and two 113's and two rifle squads from the 11th cavalry.

I was challenged for the password, and responded.

The sergeant of the Quad 50 jumped from the gun truck and ran over to my Jeep. "Kid, what the hell are you doing this time of night out here," the sergeant asked.

"Oh, just some silly shit that some doorknob dumb CIA and unit officer put me up to. I'm having a blast screwing with their minds. Hide and watch, you'll hear about it later and have a hell of a good laugh when you hear how ignorant they will look," I responded.

"Kid, I'll radio the crew at the west end of the village; they will pull the barrier and wave you on through," the sergeant assured.

"Thanks, Sarge!"

On through the checkpoint the little Jeep roared, and up to the entrance of II Field, the MP approached me and spoke, "Courier from Duster Compound?" The MP nodded and said, "Been expecting you," as he pulled open the gate. I literally strolled into II Field Headquarters with a touch of arrogance.

"Get this thing off of my wrist," I addressed the clerk.

The clerk wide eyed mouth dropping open responded, "You made it!"

"What the hell did you expect?" I barked.

A full colonel came out of the inner office and said, "You made it; well, you can spend the night here with us and go back in the daylight."

"Sir, do I have to respond to you like you are one of the 90-day wonder captains in my battalion?" I asked.

"No, I'm a West Point man and I would not want to have that label, especially in your eyes. You do what you want, when you want, where you want, but you are doing one hell of a job, trooper. I would give a piece of my retirement to have 10,000 like you in my command, but without your bar room vocabulary."

"Thank you, sir, that has to be the first compliment I've had in this insane shitty war," I said as I saluted the colonel.

The colonel said, "Kid, I have made up my mind about you and believe it fits you so well. Billy the Kid is 10% insane, 100% lunatic, fool hardy, and deathly foolish; a grenade launcher is his outlook and death does not faze him much."

My response, as any good Native American would say, was "Well colonel, in the book of life with the Great Spirit it is written when one is born, the death will come and why fear what will come."

The colonel replied, "Well you definitely have something there, son."

I walked out and returned to Duster Compound the same route and manner.

The 5/2 was just beginning and preparing for the big move to the 199th Light Brigade base. Chief Baltier had vengeance for me and had a plan that I had to die and was going to do his best to see it happen.

A radio call came in from the southwestern Central Highlands. It was barely south of the II Corps and III Corps boundary for an engine for a Duster.

In such remote locations, normally the mechanic and engine pack would be flown in. But this time, Chief Baltier had a good engine pulled by the strip crew, put it down on the 1/4-ton trailer, and boomed down on the trailer.

The next morning the chief would not allow me with a newbie mechanic to leave Duster Compound until 1000 hours. I was given directions to a new 25th Infantry firebase in terrain not familiar to me, deep into the Central Highlands jungle, above the high plane.

I went to the hooch and made four sandwiches with mayo from bread that mother Catherine had sent from Kentucky, and the canned ham from II Field PX. In one of my two canteens I put grain alcohol and Tang acquired from Captain Soltis' medics, and water in the other. I packed the sandwiches wrapped in Reynolds wrap from the Long Binh commissary.

I had a platoon member strap and tie off the beloved eleven-inch blade to my right arm that went from the wrist to the elbow. With my AR-15 battle rifle in hand and ammo bag slung, I went off to the green-line for the bag of M56 grenades. I had my personal M79 sling and two bandoliers (total 12 grenades for the M79) slung over a shoulder.

The 19-year-old mechanic and I were ready. He had an M16 rifle, one bandolier of ammunition (10 magazines, 20 rounds each) and tool-box in hand.

I had the trailer hooked up, having to air the six ply tires to 60 pounds; the tires were overloaded. "Saddle up newbie mechanic, time to take you to the real war," I said to the kid only in the country three days.

We headed up Highway 13, going to Cu Chi then to Tay Ninh. Tay Ninh was just getting started and the hard pack ended. We were driving on a mud road freshly cut through the jungle. The foothills in that area were difficult for the little Jeep. We would hydroplane through inches of thick mud in the low places and have to shift to 2nd gear to get up each hill. It was 45 mph downhill and as low as 30 mph up hill in 2nd gear.

We drove deeper and we saw a two-man Viet Cong recon team, each standing by a huge tree, literally seeming to be in shock for what they were looking at. Two fools in the middle of nowhere.

The young newbie started to panic, crawling almost into my arms screaming, "Protect me, protect me!"

With my left hand, I slapped the newbie across the face three times and screamed, "Straighten up you pantywaist pussy! Lock and load your weapon, put the safety on fire, do not put your finger on the trigger unless you must fire; fire only if fired upon. Get yourself two grenades from my bag and put them in your crotch. If you must throw the grenade, cook it one second so old Charlie can't run!"

"I'll make a fighting soldier out of you if your coward ass doesn't get us both killed. It's time for you to grow a real set of balls; you are sure going to need them in these highlands."

We had left Duster Compound at ten hundred hours and rolled into a makeshift firebase at 2030 hours. There were two Dusters, one with the armor plating removed — obviously the one needing the engine replaced.

The base was about 200 feet square with about a 25 foot kill zone cleared back.

One hell of a firefight was going on about 100 feet east of the kill zone — a solid scream and roar, a blue white glow from high explosive grenades and weapons of both sides.

One E-5 came running, grabbed my hand and shaking it. He greeted me saying, "How the hell have you been getting along Billy the Kid?"

"Like everyone else, just getting by," I answered.

"Heard you were killed down on Highway 1." The sergeant seemed surprised.

"Well, you know how things can get screwed up on the base camp with those base camp warriors. You know what I mean?" I indicated.

"Yeah, say you can stay with us tonight, we are going to kill a lot of gooks tonight, there is a battalion of VC coming. I'll set up my 60 for you to fire and get a grunt to feed and clear it, if it fails. You guys down on Duster Compound are just contending with part time farmer VC, but up here are hard core full time sons of bitches," the sergeant boasted.

"No thanks, Sarge, I've done my job getting this newbie here to change that engine. I'm headed back to Duster as quick as I can drop the dog leg on my trailer."

I unhooked all lights with exception of the right front, saddled up and rolled out headed home for Duster Compound.

Coming up on Tay Ninh, I called in, and was told to roll on.

Coming upon Cu Chi I radioed, "Cu Chi, Cu Chi, this is Duster Chief, I say again Duster Chief. Cease fire, hold your fire, I'm coming through!"

In the background I could hear a second voice. "Who the hell is Duster Chief?"

I responded, "Damn fool, this is Billy the Kid, I say again Billy the Kid; don't shoot me, I'm headed back home!"

"What the hell is that lunatic doing out this time of night," a voice roared.

"Doing my job base camp warrior, doing my job!" I replied.

I rolled into Duster Compound at 0430 hours.

CHAPTER TWENTY

Captain Moore

Now this new Captain Moore was the dumbest individual in the entire theatre, no doubt. He and a new African American sergeant first class would be transported by me to Cu Chi to meet the battery commander.

Captain Moore sure was proud he had made captain. He wore the chrome bars on his collar and boonie hat. He saluted first and saluted everything that moved, and expected a return salute from whoever or whatever.

The trip began uneventful; however, every vehicle on Highway 13 got a salute from Captain Moore.

I spoke to the captain, "Sir, if I were you, I don't believe I would be saluting out here."

"I'm an officer, the troops must salute me," he responded.

North of Phu Loi, the captain saluted a 10-ton dump truck.

The driver hung his head out and screamed, "Stick that chrome up your ass, asshole!"

Captain Moore asked, "What was that he just said?"

I answered, "Well sir, I think he told you to put that chrome up your ass."

After about two minutes, several shots rang out from about 250 yards. Rounds from an SKS, no doubt, whizzed through the Jeep.

Captain Moore pulled out his .45 and laid it on the bench. He then rested it on the rear of my shoulder and said, "Hold still." The fool was going to return fire at 250 yard with a .45. The muzzle blast could do serious injury to my head.

I did exactly what Chief Silvers had taught me. I leaned out the door opening and made myself as small as possible. For about a mile I drove like that.

Captain Moore asked, "Why didn't you hold still I could have gotten a shot off."

I did not respond. The thought in my head, "Damned fool, the .45 is only good for 50 yards in the hands of a well-trained, experienced, intelligent trooper. This door knob dumb captain was none of the three."

Upon returning to the base camp I announced to Captain Moore, Chief Baltier and the sergeant major, "I'm not driving any more; you idiots are going to get me killed with your dumb ignorant stupidity!"

Chief Baltier barked, "You will drive that machine, and that is a direct order!"

I made a request, "Permission to see Dr. Soltis, please."

"Sure, no problem," they all agreed.

I walked the 100 feet or so, and entered the battalion clinic.

"Captain Soltis, pull my medical file please," I requested.

"Sure," the captain replied.

With medical file in hand, I pointed out to the captain, a non-combat profile, non-combat zone, no combat arms, no military driver's license and other restrictions, dated March 28, 1969 at the induction station, Louisville, Kentucky.

I demanded to the captain, "Pull my driver's license; they are trying to get me killed!"

It stated, no war zone, no combat MOS, non-combat arms units, etc. I flipped the military driver's license onto the counter and stated, "I have a fool doing his dead level best to get me killed, void the license!"

Dr. Soltis voided my military license. He then informed the officers that my license had been voided.

The next morning a staff of field grade officers led by a full colonel and accompanied by a Quan Loi police officer were putting all troops at attention. The group was accompanied by a 14-year-old local female. She pointed at me speaking to the Quan Loi police officer.

The colonel spoke, "Trooper, you are now under house arrest until an impartial investigation is completed. You could possibly be turned over to the Saigon government for trial in the courts."

The guard was immediately placed on me, but no weapons were taken away from me. I still had in my possession the AR-15, the M16 and the M79 and 401 rounds 5.56. For two days, senior NCO's and captains who did not like me would walk through the hooch and harass me.

One captain made this statement, "Hey, the Saigon government is taking you to a civil trial, convict you and hang you with a rope until you are dead, dead, dead. Got that!"

I did not respond.

One officer came in and sat down and explained some about military law and civil law of the Saigon government. He said, "This 14-year-old is the only witness; if you go to trial and judged innocent, the courts and justice system will automatically hang her for perjury, thought it would help you to know that."

The third day no guard showed up to guard me.

I asked a senior NCO, "What is the deal, no guard?"

"The charges have been dropped for lack of evidence," the NCO answered.

Later that day I questioned two Rangers from the 30th, a sergeant first class, and a master sergeant who were interpreters and advisors to Head & Head 5/2.

"It is okay now, you will not have the charges brought back," the master sergeant assured him.

The sergeant first class added, "The young man and his sister were Viet Cong. We had been watching them for some weeks before he climbed onto the truck."

"What happened to her, why is she not going to testify?" I asked.

"You do not need to know, go home to Kentucky, forget the terrible experience of this war of my country, find a good woman, have children and make a good life," the Ranger said.

"Well, she may change her mind," I said in a worried manner.

"No that is impossible, we went to the village and took her prisoner. We took her to a rubber tree plantation. She dug her own grave, a deep one, we both shot her in the head at the same time. It was the orders of Lieutenant Colonel Do Van Dien; remember, he is the man that gave the award of "the land of the tiger" to you on Duster Compound for killing those six Viet Cong and wounding six more when it was only you." The master sergeant admitted, it is his way to repay you for the uncommon valor of the Kentucky hillbilly."

"We know of no American that is like you and what you have done for this country," the Ranger added.

The two Rangers merely turned and walked away.

The major came to see me that same day. He said, "Seeing as how you are no longer the driver and courier, we are reassigning you to the night crew still holding Duster Compound. You will not spend your days here on the base camp. You will be picked up each morning on Duster by Specialist Shookman and be one of the three that will bring in the troops from firebase of Alpha, Bravo, and Charlie batteries, so these troopers can return to stateside duty and some with less than six months until they ETS from this man's Army."

The major continued, "Gather your field gear and report to the captain in charge of the night crew at 1700 hours to return to Duster each night. You will be dropped off each evening in time for greenline each night."

I responded, "Sir, yes sir," and saluted the major.

The thought had crossed my mind that this was better than being screwed with by these 90 day wonders."

So, it was Big Shook, me and the man who had changed the Duster engine in the Central Highlands. We were now the recovery crew bringing out the few remaining troopers still alive and their bodies still intact to send them back to the World.

However, one Quad 50 from D71st ran escort through the whole ordeal, when that "Whispering Death," as the Viet Cong had named the machine, fired for about 90 seconds or so. One of the Duster crew commanders said it best: "Man, when the Quad is finished, it is so quiet you can hear Mighty Mouse whizzing on a ball of cotton."

The 24 men holding Duster Compound were all assigned permanent bunkers, as well as the tower crew. The captain assigned me, and his two best and seasoned troopers, to the bunker 3 Alpha, as next to the gate was the weak link in greenline defense. Since there was no longer a Quad 50 to cover the other side of the gate, the best and most experienced observers were in the tower.

The captain was doing a fine job of command. His staff sergeant of 32 years was on his 3rd tour here on Duster Compound.

It had gotten quiet. Charlie was no longer probing with recon. It was as if the communists had forgotten our compound. Chief Silvers had taught the platoon, "When there is a total lull, get ready because the VC will hit you with everything they can muster and then some."

Mother Nature had returned. The little deer in small herds had returned and would graze in the minefields at night. This assured the men that as long as they were present, Charlie was not near.

Early one morning about 0200 hours it was quiet with the deer grazing in front of 3 Alpha. The E-5 was on watch sitting behind the M2.

The rather silly young man who had referred to the Quad 50 as "that double barrel electric corn sheller," was sleeping on the ground next to me when all of a sudden, the trooper behind the M2 cut loose the loudest fart ever released by any human.

The trooper next to me popped up and loudly shouted, "Who fired those shots, that was not authorized!"

The three of us had one hell of a laugh. The small herd of deer ceased grazing and were staring at the gunner, as if to agree with the statement.

Some idiot brass had handed over all the ammunition trucks of the 5/2 to the ARVN military. The only vehicle to bring in the troops and transport the infrared over white light, searchlight teams of the I-29th searchlight teams and Duster crewmembers of A, B, and C batteries being bought in, was the 10-ton wrecker, with fifth wheel and the 40-foot low boy.

The average crew brought in from each firebase included Duster crewmembers and the two-man crew of the searchlights. The searchlight was created in a prepared wooden crate and nailed down to the wooden bed of the low boy.

The very last crew to be brought out was loading up. The crated searchlight was nailed to the wooden bed of the low boy, then crewmembers with one inch rope secured around their waists were tied down to lanyard rings normally used to tie down the Dusters.

An argument broke out between one of the Vietnamese troopers and a crew chief sergeant over the M60 that was normally part of the tactical weaponry for the Duster. The American sergeant loudly said, "Papa San, you can come to the 25th Infantry base camp for this 60. It goes in with me, got it!"

Everyone was set, one sergeant loudly shouted, "Big Shook, roll this machine, we are ready to return to the World, I just hope we all get there alive!"

Big Shook rolled over and fired up the massive turbo diesel engine and rolled out the last time with the last crew of the three batteries standing down. There were only 16 Dusters remaining with D Battery based out of Binh Hoi, and only 18 Quad 50s remained with very high combat losses.

A few miles out, all hell broke loose. We were taking multiple mortar rounds from at least three mortar tubes. The rounds were falling short. We were taking rounds from about 400 yards from the AK-47s and SKS weapons. I sat in the center of the truck cab, by far the smallest of the three men, and the 19-year-old mechanic sprang to his feet and began firing the 60, screaming "Feed me, feed me, mama, mama, feed me," with tears streaming from his eyes as was normal for him in fire fight.

Big Shook down shifted one gear to peg the tack then back to fifth gear pouring on all the diesel fuel possible. I was actually on my knees directly under the 60. One round fired from the M79 max range. I cursed out of rage, with my left hand reaching around to feed the 60, dropping the 79, then grabbing the AR-15, which was locked bolt to the rear and safety had been already placed on full auto. One push of a button, so to speak, well at least one magazine fired directly into the bush right handed and one handed. Dropping the AR-15, I continued to feed the 60, with hot brass going down my collar and the mechanic dancing around in the excitement all over the legs and feet of my body with his boots.

The Quad 50 slowed to about 20 MPH, raking the wood line and bush with 2400 rounds a minute. After about ninety seconds, the driver of the Quad 50 poured on all he had to catch up with the 10-ton and the low boy.

No one spoke. Big Shook just looked at the two men and smiled the usual. Actually, he was a rather quiet and kind person.

The 10-ton pulled into the 25th Infantry base camp.

"Shook, let me grab a clean set of fatigues. I can smell myself coming ten feet away after a week in this set. I'll be ready for greenline tonight on Duster," I admitted. All I had left clean was one of my two sets of tiger stripe uniforms.

Big Shook rushed me to compound for what would be the last night for the night crew, as of ten hundred hours tomorrow, the 30th ARVN Rangers would assume full commitment for the compound.

I quickly shaved, showered and went to report for the guard mount for the last time on Duster Compound.

The only two Dusters were in the southeast and southwest corners. There was no backup equipment, no Quad 50's.

The ammunition dump had been pulled. The red status bunkers had been cleared out of ammunition and grenades. Only the normal six bunkers still had ammo. There were only 1200 rounds for the M2's.

CHAPTER TWENTY-ONE

Last Battle

The captain who had commanded the night crew for six weeks announced, "Gentlemen, gather around me. There will be no guard mount tonight. If you have not cleaned your weapon, do so hurriedly before sun down. The staff sergeant has a cleaning kit for you to use."

"Gentlemen, I was drafted as most of you were. I was a school teacher with a math degree, teaching high school math, before going to Officer Candidate School. Gentlemen, we are considered the least desirable and most expendable part of the 5/2. That is why we are the night crew. Gentlemen, Hanoi has 600 NVA handpicked and selected who have been traveling by night for the past six weeks and hiding in tunnels by day. They have 40 Viet Cong with shoulder fired rockets that will shoot down any helicopter, and they have heavy machine guns comparable to your M2 50's.

"Puff the Magic Dragon from Danang is being brought here, but if it gets bad up there they will return it back up north. We have been ordered to hold this post and position at all costs, to the last man whatever it takes."

A number of the young men began to cry.

I had been sitting on sandbags honing the 11-inch blade on the leather on the rear of my jungle boot.

I stood up, combat knife in hand, thumb down to the earth, rather level and hard then said, "Mama is not here tonight, mama cannot help you; only two things might get you through tonight. That's God and your military training!"

"That's the spirit," the captain blurted out.

"Aren't you afraid you will die tonight," one of the young men asked me with a shaky broken voice.

"Yes, but I'm going to take every one of those commie sons of bitches with me if I can. I intend to pile them up like cordwood," I barked out.

The two ARVN interpreters, the sergeant first class, and the master sergeant were there with the platoon. The master sergeant spotted movement on the more or less deserted compound.

He made a quick request to the captain, "Sir, there is a problem; your .45 quickly!"

The master sergeant Ranger double timed to the south of the compound and caught a 14-year-old Viet Cong trying to exit the compound through the minefield.

The master sergeant brought his prisoner back, hands tied in the front with a stick run through his arms and behind his body; he returned the .45 to the captain.

The master sergeant commented, "I need a Zippo lighter, a cigar, and I have my combat knife. I will interrogate this Viet Cong prisoner, but you are not allowed to watch."

The master sergeant pulled from his pocket some folded papers and said, "Captain, these are maps of the entire compound that this Viet Cong has drawn. I will be only a few minutes to interrogate him in this connex compartment."

The platoon could hear a body being thrown into the steel walls inside the compartment; there were blood curdling screams of pain.

In about five minutes the master sergeant exited with the prisoner. There were burn marks about his body from the Zippo and the cigar. Both arms were broken above the elbow. There were long shallow cut marks about his body, and salt particles around the wounds from c-ration salt packets.

The master sergeant commented, "He has told me everything. Soon there will be the normal trial and he will face the normal firing squad for treason to my country."

At this time the captain made the statement, "Gentleman, it's my call, and I have declared and called in to MACV "Broken Arrow" fire inside the battle area."

"The call for volunteers to help us is 100 3/4 Cav from the 11th Armored Cavalry. They are now dropping about 200 yards out from the west wall. These 100 men are all the help we will receive."

At that time the A Team commander approached the captain, saluted and said, "My men and I can do it for you, sir."

The salute was returned with the response, "Thank you sir, my men and I greatly appreciate what you are doing."

The A Team commander left and in a short time brought the entire A Team to us. Each Green Beret took out a note pad that came in the Red Cross S+P packs to write a quick letter home.

A sergeant standing next to me started his letter to his wife and two children back home. He held a photo of himself in the Class A greens sitting next to his beautiful wife. In the photo, he had his son about one year of age sitting on his lap. His wife had their little girl of about 3-years-old on her lap. The letter began:

"My darling and loving wife,

Things are starting to look bad here.

If I do not make it, teach our children what their daddy has done for our country, the land of the free and the home of the brave. Hug and kiss our babies for me and tell them that daddy loves them.

When they are older help them remember me.

With all my love."

The 12 men of the A Team each took out a purple bag that Crown Royal whiskey is packaged in and put the letters, their watches, rings and photos after each man kissed their own photo. They put all in the bags and handed them over to our captain for what one might assume safe keeping.

The A Team went back to their side and shortly returned geared up with battle weapons. They broke into four teams of three men. They were heavily laden with 100 pounds or more. One M60 per team, M60 100 round belts around their necks piled to about ear level, grenades in bags hanging on each side, and one man of each team wearing a claymore vest full front and rear.

Each team walked over the berm, and set up just outside the berm beginning with the setting up their first line of defense with claymores and high explosives, commonly called automatic ambush.

With the 11th Cav 3/4 Cavalry set up in position with 33 ambush teams in place, the A Team walked over the wall where 50 to 100 men should have been. The captain said, "Gentlemen, fix bayonets, prepare to die for God, country and each other!"

As the men started to deploy, the captain asked me to stand fast for a minute. After the men were out of earshot he said, "I was losing control of the platoon, thanks for bringing the platoon together. I'm sorry for issuing you the egg meal in the c-rations for each meal, please forgive me."

"Captain, there is nothing to forgive, a meal is a meal. I kind of got used to the eggs, ketchup sure helped some," I replied.

"Thanks," the captain said as he walked away to join the staff sergeant at the command post. About thirty feet in his walk the captain stopped and turned to me, "See you in Hell if we don't make it!"

"We will split the gates of Hell together sir, as a team." I replied.

"Right on kid," the captain blurted out, and walked on to his command post.

The three men climbed the ladder to take position in the 75-foot tower. The six fire teams manned their bunkers laying what magazines out in front of them for quick reload.

I took my position on 3 Alpha next to the entrance gate under the overhang of the bunker. I laid out my poncho, copying what I had observed on firebases of the 25th Infantry. Put that poncho down to help keep dirt and debris out of those magazines. A trooper sure doesn't want any hang fires or jams on a battle rifle in the middle of a firefight.

All of the platoon were ready mentally and physically to spring up and go to bayonet, combat knife and hand to hand with Hanoi's chosen battalion.

The sun fell from the sky then dusk. At 2230 hours, the first firefight broke in the southwest corner where the 3/4 Cav engaged the first team of commies.

It lasted about 15 minutes. There was a solid scream and roar of a combination of machine gun, battle rifle and high explosive weapons. A solid mushroom glow of pale blue engulfed the fire teams of the 3/4 Cav that fought the first engagement that horrible night.

Two Puff the Magic Dragons rotated and actually would meet each other one going out and one going in to refuel and rearm.

About midnight, the first of Special Forces fire team engaged Hanoi's finest. Such was a repeat performance of the 3/4 Cav. A number of engagements raged through the night.

At 0430 hours, two engagements broke out at the same time, one directly in front of 3 Alpha in front of my bunker, the other in front of 2 Alpha in the southwest corner. The position held by 5th Special Forces and the A Team.

Rounds and fragment of high explosive weapons were finding their way home and coming into the upper open areas of both bunkers. Bunker fire teams were as low as possible in the bunkers with one man in each bunker, only a steel pot showing with a pair of eyes observing and preparing to be over run.

Again, about 15 minutes of time lapsed and all was quiet.

About three minutes later, one lone NVA with fixed bayonet, made it over the berm at 2 Alpha. 2 Alpha position had fired dry. One man of the fire team came out of the bunker and went one on one with the NVA who obviously had also fired dry.

The end result was one dead NVA and a wounded 2 Alpha fire team member.

First light came. UH-1C medical evacuation helicopters came screaming in. The first man loaded was winched upon a litter from the west wall. His right leg was missing six inches below the knee. The medical Hueys were running hard to get the wounded out while they were still alive.

On our south berm, near the two bunkers, we had four of six men down but still alive. I looked at the gun position 2 Alpha.

The one man left standing was working on his brother to keep him alive.

The captain pulled ammunition from 3 Alpha and an opposite bunker on the east wall because 2 Alpha had gone dry.

Two men were sent to the south wall because those two bunkers had only one man each still standing unscarred. The captain came to 3 Alpha to speak to me, "Hell of a night kid!"

"Yes sir, one I'll never forget if I live to be a hundred years old," I answered.

"Well kid, we've got 67 of the 100 Cavalry men are either dead or out of the Army because they have lost limbs. All 100 will draw a Purple Heart. Six were killed outright, and three bled to death. The other three will make it, but one might lose a leg. My four men will live but their last place will be Walter Reed in D.C. The war is over for all but twenty-two of us left unscarred."

The captain stopped speaking and the men just watched as three men, arm and arm, were barely walking to get about 30 feet to the last medical evac helicopter.

Three men were wearing green berets. The one in the center had the leg of his battle dress trousers cut away with Ace bandages from the crotch to the ankle. He could not walk on that leg. Each man on his right and left were holding him up. He was walking best he could on the remaining good leg. They were, and are to this day, a special breed of men. They are set aside from all other military forces. The three men were just laid aboard the helicopter as the medics began feverishly to keep them alive. They were rushed full throttle to Long Binh Hospital.

That was the last of the wounded. Now began the chore of removing the dead, to take them to graves registration for the slow journey back home. The last to go were the nine Special Forces.

The captain and I watched silently as the last man was being picked up. The two medics very gently, with the greatest respect put a Green Beret on a litter, placing his hands together just above the belt line. They picked up and placed his beret on his chest. The two medics were wiping tears from their eyes on blouse sleeves as they worked. The litter was carefully placed in the helicopter. As the helicopter lifted off, it came to mind one part of Sergeant Sadler's *Ballad of the Green Berets.*

A young wife waits
A Green Beret has met his fate
A Green Beret makes his last request
Our young son will be a man some day
Give our son one chance to win my green beret

The captain spoke, "Well kid, this is what it's all about, stopping the communists before they wind up taking over the entire world."

"It sure was a hell of a price to pay, captain sir," I replied.

I was placed in the guard shack until a team from II Field, led by a general, inspected the base.

At ten hundred hours, I was the last American solider to climb onto a deuce and a half headed eight miles away to the 25th Infantry base camp.

Young men were cheering, some crying as the lone deuce and half pulled into the battery area. As I walked towards the hooch one of the men tried to hug me. I pushed the man back, screaming, "Don't touch me, it's not over yet, the commie bastards killed my brothers on the A Team. They've still got Hell to pay!"

"Leave the kid alone, give him some time by himself, trooper," one of the captains commented.

Soon Christmas would be here. It was time to mourn the dead. I had thoughts of children who had lost their fathers who had given their lives in the name of freedom for the oppressed of the world, children who would grow up with only a photo of their father in uniform on the mantle — a man who had earned his green beret. America's finest, America's best.

The young wife who waits back home will look up and see a chaplain and an officer coming to her door to tell her that her man has already began the slow trip home. Then he will be placed in his final resting place in hallowed ground for eternity.

CHAPTER TWENTY-TWO

Christmas 1970

All of the men in Head & Head were excited about Christmas just around the corner. My mother had sent what the platoon named "Charlie Brown's Christmas Tree," with lights. Of course, the lights were not allowed because of blackout regulations. They were not available at the large PX in Long Binh. The men of the platoon, not on the night platoon, had put up the little Christmas tree and hung a "Santa" from the ceiling of the hooch on the 25th Infantry base camp.

All of them were talking about Bob Hope and the beautiful ladies coming to Long Binh Christmas Day and all possible wanted to be present; that is, those still alive.

People were coming from Long Binh, II Field to Binh Hoi. They just wanted to see my little Charlie Brown's Christmas tree setting on the little refrigerator. With its eight little lights burning bright, much like the Star of Bethlehem the night Jesus Christ was born, one of the ladies from the Trailer City commented. Some thought I was the lunatic, and they wanted to meet in person.

A two-star general walked in from II Field. I jumped to my feet from the cot I was resting on. "At ease trooper, relax. I've heard about this little Christmas tree with lights. I just had to see for myself." After admiring it he asked, "Where did you get it, where did it come from?"

"My mother in Kentucky, just south of Knox, sent it. She wanted the platoon to have a touch of home here in this Hellhole on the special occasion sir," I answered.

"Well, you write your mother that I said she is one very special mother."

"Sir, yes sir! I will tell her you said so, sir," I replied.

"Sir forgive me, would you like a drink, vodka, crown royal, a bourbon from Kentucky, a Jim Beam or something sir?"

"No thank you, son, but thank you for asking. It looks as if you have a touch of your thoughtful mother in you," the general commented.

"Well sir, I give it my best sir," I answered.

With that, the general just smiled and walked on through the hooch looking at what the platoon had put up for the holiday.

Everyone celebrated the holiday. Many of the battery attended the Bob Hope show, with hangovers of course. The day after Christmas, there was a meeting of officers and senior NCOs.

With only D Battery remaining and troops returning to the States, many positions were eliminated. Such was true of the position I had held as courier and driver.

One officer suggested I be transferred to the 25th Infantry to become a tunnel rat.

Sergeant Timolsic stepped forward and said, "He has less than 90 days to ETS. You cannot transfer him out of this battalion. I need help in my battalion welding shop, I need him. He has some background with oxy/acetylene in his technical school before the military."

The lieutenant colonel replied, "Okay, but he keeps his mouth shut, and keeps his distance from Chief Baltier."

Timolsic replied, "I'll see to that sir, he belongs to me now, thank you sir!"

And so, it was, I would now sharpen skills that would be valuable when I returned to Kentucky. I began the training to weld up trigger bars for the Quad 50. Heating them to about 1200 degrees Fahrenheit after the welding process and quenching them in linseed oil to temper them. I heat treated them three times each.

I began welding holes in Duster fuel tanks with the ARC welder — holes made by the soviet heavy machine gun.

CHAPTER TWENTY-THREE

Angels

Mid-February, twelve men, one captain, one staff sergeant and ten enlisted men, including myself, became very sick. Our throats swelled so bad chairs had to be inverted to give us about a 30-degree elevation to keep the wind pipe from closing off and keep from suffocating.

We reported to the battalion medical clinic. Eleven other men and myself were quarantined. A red flag went on my orders, until it would be determined whatever was causing the illness was not contagious, and or curable.

Dr. Soltis informed all twelve patients that the viral infection was contagious. It was not curable and we had two choices. We could choose to be written off "missing in action" and transported to what was called Christmas Island. Air drops made every day to Christmas, whatever a man requested by radio, he would receive in an air drop the following day.

The second choice was 3/4 Cavalry sick squad.

They lived daylight hours in the corner of the base camp, quarantined in the restricted fenced off area. They literally had little houses, cooking facilities and such. Local females, all volunteers, lived with them, cooked and cleaned.

Any alcohol, or controlled substance was furnished, as they would never be allowed to re-enter the United States.

At 1900 hours, they would report to the helipad for night ambush three-man teams. I was informed that if what I had was contagious, that there was no cure and I had those two choices. It would be long days until the medical results would return from a lab in Japan with the results. I made a choice almost instantly. If the results were that I could not have reentry back to the States. I would join the 3/4 Cav and kill more communists.

The results came in, unknown viral infection, non-contagious. Twelve good men were all relieved. The current NCO, at that time, relieved me from my duty due to the seriousness of the illness.

The sergeant said, "Stay in the battery area, do not leave this base camp. There is a reason for this."

"Okay, Sarge," I agreed.

At this time, an old friend came in from Firebase Blue. It was the preacher. He was from down about Mayfield, Kentucky or maybe somewhere in western Kentucky. Our paths had crossed several times since we had met in Cambodia.

"I just want to be with a man that is home folk," he explained. He began to tell of his childhood and his parents. How they met in the early '40s, fell in love and married. He was the only child. His father had a heart attack while off bearing the saw on the sawmill. The owner of the sawmill allowed his mother and him to continue to live in the little shotgun, sawmill lumber house. He told that mama would be given old sweaters and such. She would take them apart and knit booties and sweaters for him when he was a baby.

He grew up wearing hand me down clothes and shoes from schoolmates; his mama would cut cardboard and put in his shoes to cover the holes.

Children would tease and make fun of him, "My mama gave you that shirt, it used to be mine."

He was on the free lunch program at school almost never receiving a piece of meat, just extra beans.

His parents had never owned an automobile. They walked everywhere and sometimes someone would give them a ride. The preacher of their mom and dad's church would take them to the doctor if needed.

He told about kids in school harassing him and looking down on him.

He told about begging his mother on his sixteenth birthday to sign for him to join the United States Army. Mom finally caved in and signed for his seventeenth birthday. He did basic training then off to Fort Sill, Oklahoma. He earned the 17B20 military occupational skill, Dusters and Quads, a gunnery crewmember.

He was a hold over at Sill as an assistant Duster instructor until his 18th birthday. Then orders from Department of the Army came for the 5/2 and landed him in DELTA Battery out of Binh Hoi.

He then began to speak of his religious beliefs and of the Holiness church he was brought up in. "An angel from heaven came down and told me to pray some more, get right with God and Jesus Christ as I would soon die."

I responded, "Aw man, you have the short time jitters. Relax man, ten hundred hours tomorrow you will come in for good, and then back to the World."

The sergeant major stared in the doorway of the hooch addressing the preacher, "Trooper, you better get your ass on the helipad to head back to firebase or you will be Article 15'd. Now Trooper!"

The preacher shook his head, "See you in Heaven someday kid," then walked out.

At about 0030 hours (12:30 am) the battalion clerk ran through each hooch screaming. Firebase Blue has been over run, fire in the battle area, Broken Arrow!"

A few minutes later, Captain Moore came walking in and addressed me, with M60 in one hand, and an M79 in the other. Captain had three belts of rounds for the M60 draped around his neck, 100 rounds each belt, and two belts for the M79 and a map case with about six or so M56 grenades hanging from his shoulder.

"Get up, off of that cot! You are driving me to Firebase Blue. They need help," he ordered.

For a minute or so I just stared at the hero wanna-be. Then I responded, "Sir, I have no driver's license, remember Doc revoked it."

"You don't need a license where we are going. I gave you a direct order, now move!" Captain Moore screamed.

"Sir, with all due respect for what you have on your collar, just screw you, and you can go to Hell for that matter sir," I shouted where everyone could hear.

The sergeant major intervened, "Captain Moore, go put your weapons away, the base commander got them in this situation and they will have to get themselves out. We will send help at first light. If the lieutenant colonel has to lock your heels, so be it. You can be arrested for disobeying a direct order. Do I need to get the lieutenant colonel?

"No sergeant major," Captain Moore responded.

The sergeant major walked over and sat down on the foot of my cot. "Son, you don't have much time left. The lieutenant colonel and I have been running interference for you for a long time; you can't keep talking to these battery grad officers like you just did. It will cease now." The sergeant major added, "Keep your mouth shut and stay out of sight. Do you understand me?"

"Yes, sergeant major, it won't happen again," I answered.

Everything was solemn and quiet in the hooch.

At about 0430 hours the sergeant major returned to the hooch. "Kid, do you think you can positively identify the remains of your buddy, the preacher?"

"Sergeant major, he was much more than my buddy, he was my friend. Yes, I will do my best to do whatever needs to be done," I answered in sadness.

Vehicles transported the A Team to Long Binh to identify bodies. The sergeant major led me to twelve body bags neatly lined up. There were so many others, however the men of the 5/2 were separate.

The sergeant major unzipped one of the twelve bags, exposing the upper half of the body of the preacher. He looked so cold, and had a complexion of grey with a strained look on his face, the look of death that no doubt had bled out. All, was quiet and still.

On my knees, I reached down and took the preacher in my arms, and held him. In silent thoughts, I said my goodbyes. Then kissed my friend on the forehead. I laid the preacher down ever so gently. I felt the lower part of the body bag. His legs were gone about six inches below the knees, and could tell there were still tourniquets on both legs.

"Is this him? You can positively sign this document stating this is him," the sergeant major asked quietly.

"Yes," with pen in hand, I put my signature to the document. I stood up and walked away noting two bodies not in body bags. I approached with curiosity, "My God, that is a uniform, class A dress of the Soviet Union," I thought he was probably a Spetsnaz.

The large I.D. card was pinned to the uniform. A photo of his wife and approximated two-year-old daughter sitting on the lap of the blonde haired and blue eyed mother in the color photo.

About a minute later, an officer of the CIA literally dragged me away from the Soviet trooper and a tall Asian in black pajamas. "Son, you will have to keep your mouth shut about the man you just saw, if word gets out it could start World War III."

Twelve of sixteen good men were lost from the 5/2 that night. Only four walking wounded would return to the World.

Dear Family,
February 1971

In the very near future, the undersigned individual will once more be in your midst, dehydrated but happy, to resume his place in "The

World." In making your joyous preparations to welcome him back into organized society, you might take certain steps to make allowances for the tropical words, he might be a little Asiatic from the Vietnamesitis and overseasitis and should be handled with extreme care. A little time in the "land of the big PX" will cure this, however.

Therefore, show no surprise if he looks around for his steel pot when offered a chair, insists on carrying his own tin can to the dinner table, or wakes you up in the middle of the night for guard duty. Keep cool when he pours gravy on his dessert or mixes peaches with his mashed potatoes. Take it with a smile when he insists on digging up the garden to fill sandbags for the bunker he is building. Be tolerant when he takes his blanket off the bed and puts it on the floor to sleep on.

Abstain from saying anything about powdered eggs, dehydrated potatoes, fried rice, malaria pills, fresh milk, or ice cream. Do not be alarmed if he should jump up from the dinner table and rush to the garbage can to wash his dish with a brush. After all, this has been his standard. Also, if it should start to rain, pay no attention to him if he grabs a towel and a bar of soap and runs outdoors for a shower.

When in his daily conversation, he utters such things as "Xin Loi", "Choi oi", "te te", "bo ku", and "same-same", just be patient. Simply stay calm if by some chance, he utters "Numbah 10", because it means nothing more than "Bad News". Do not let his colorful descriptive vocabulary surprise you, and don't let it shake you if he picks up the phone and yells, "Saigon, sir", or says "Roger, out" for good-bye or simply shouts "Working!"

Never ask why the Jones's son held a higher rank than he did, and by no means mention the term "extend." Pretend not to notice if at a restaurant, he calls the waitress "mama-san" or uses his hat for an ashtray. He will probably keep listening for "Homeward Bound" to sound off over AFVN, and will be baffled by the choice of radio and T.V. stations. If he does, comfort him, for he is still reminiscing.

Above all, keep in mind that beneath that tanned and rugged exterior there is a heart of gold. Treat him with kindness, tolerance, and tender loving care, and you will soon find him to be the same happy-go-lucky guy you knew and loved.

Last, but by no means least, send no more mail to the APO, fill the refrigerator, get the civvies out of mothballs, and fill the car with gas....

Because the Kid is coming home!
10% insane 100% lunatic

Jinxie talked me into going to the village to say goodbye to Na Tae (no land in the French language) and one older lady Maw. It was customary in Head and Head Battery to do so.

Jinxie asked me to take my AR-15. "I don't normally go anywhere without it or my blade," referring to my eleven-inch custom made combat knife.

So, it was to the PX for two boxes of Tide laundry soap, two multiple packs of the large bars of Ivory bath soap and four cartons of Marlboro Reds and Salem cigarettes. The cigarettes were taped to the legs, hidden, of course. The soap and Tide were hidden under sand bags under the rear seat of the Jeep.

Through the gate, out to Highway 1, on to Maw's one room little house with a porch. The one room had a full size old brass bed, one bureau, blue in color with flowers painted on the drawers. One light socket hung from the ceiling, and one electrical outlet, and there was a grill on the porch for cooking.

It was only a short time until a firefight broke out between the village protectors (their national wounded warriors no longer able to serve in Saigon's normal military).

A child, Maw's grandson, came running through and screamed, "The VC want to kill the two Americans; they know that one is the Ranger Billy the Kid."

The firefight grew closer, one man walking with a crude prosthetic right leg, was blazing away with an M1A carbine on full auto, two animal skin bags hung from his shoulders. He would rotate throwing M56 grenades and firing. Maw took both of us into the one-room house. A hidden air raid shelter behind the bed would suffice for cover and concealment.

I asked Jinxie for ammo. One eight round magazine was all he had on him. There was ammo in the Jeep, but it sat ten yards away.

I started to low crawl out of the shelter, when Maw put her hand on my left shoulder and whispered softly, "No, no, VC will kill you, chop off head, put on pole and walk through village, will say I killed the Ranger Billy the Kid."

We remained in the shelter for about 15 minutes, when Maw's grandson came running out and screamed, "VC pushed back, go now quickly."

We made a mad run for the Jeep and escaped.

Once back on the base camp I said, "When I first arrived here you were screwing off up at Cu Chi, you have blundered a number of times, now I'm going to discipline you for not having ammo on you, you said you would have plenty.

"Take off those sun glasses," I demanded.

Jinxie shook his head no.

I did my one-two-one, three blows to his face. He was bruised, bloody nose and the sunglasses were broken and hit the ground. Jinxie took some steps back and said, "Enough, uncle, I'm sorry I screwed up."

"Okay, fella, if anyone asks, you fell here in motor service, you would not want the rest of the platoon to know how you screwed up and could have gotten both of us killed," I shouted.

"Okay, okay, it's a deal," Jinxie agreed.

"Man, you are on career status, if you want to someday retire, you have to get your shit wired tight or you will never make that retirement," I said.

It was just about ten days before my estimated time served date when I caught a walking guard mount around a snack bar belonging to the 25th Infantry, first swing 1900 to 2100 hours. I walked the guard mount a few minutes playing around telling the troops, "Golly, I'm Gomer Pyle USMC." The passing troops were laughing and enjoying the clown act.

After a few minutes, I went in and sat at a table with the grunts listening to conversation. A few minutes later, a second lieutenant walked into the snack bar and addressed me. "Are you the trooper who is supposed to be walking the guard mount around this snack bar?"

"Could be," I answered.

"You better get your ass out there on that guard mount, trooper!" the lieutenant ordered.

The troops of the 25th, fresh from the bush and on stand down, began to haggle the newbie ninety-day wonder.

"Kill the butter bar," came from the corner of the room.

"Beat his ass," cried another.

"Shoot the son of a bitch," came from across the room.

The lieutenant hurriedly left and I returned to the guard mount.

At about 0230 hours, I was walking my second swing and stopped, ducked into an air raid shelter to puff on a non-filtered Pall Mall.

Well, wouldn't you just know it, here comes the newbie lieutenant, walking. He was walking the guard mount in reverse direction.

I fixed and attached the bayonet on my AR-15 stepped out and screamed, "Halt, who goes there?"

The lieutenant came on. I charged a round in the chamber and snapped the safety to semi-auto fire position and screamed, "I say again, halt, or I will drop you dead as Hell!"

The lieutenant stopped, "You would shoot me?" he asked.

"In a heartbeat," I yelled.

"I'm going to check your weapon," the lieutenant said.

"Not hardly, no one touches my gook slayer but me," I refused.

The lieutenant backed away.

Within 72 hours, a letter came down from MACV recommending punishment under the uniform code of military justice.

The much older first sergeant called me into his office. "Our young new 90-day wonder captain wants to give you an Article 15 for the stunt with that new lieutenant, the one that was the officer of the day for MACV. Here is the Article 15 you are supposed to sign, oops it just got lost in this mountain of paperwork," he said as he put it under a stack of paperwork. "You have done a hell of a job out there. I have arranged with the lieutenant colonel for you to take a six-day educational early out."

"Master Sergeant Johnson will inform you tomorrow morning, clean this place and go home, and say nothing of what I just told you," the first sergeant asked.

Early the next morning, the master sergeant woke me early saying, "Get your ass out of that rack, you are clearing post today, and leaving for the 90th Replacement. That young lady who washes clothes in the Cu Chi area, the one you were giving piasters to for written information, well, the Viet Cong took a nice photo of the two of you. Well, they raided the place and found a roll of piasters in a uniform that would not have fit you. It took her three days to die. She gave you up. Moscow, Peking and Hanoi had each put up five hundred dollars for your young ass dead or alive. Go home and crawl in a hole and pull the hole in after you. Stay away from military posts and government agencies. There, no doubt, will be a Soviet Special Forces operative inserted into the U.S. to kill you; you've broken about every rule of warfare in the Geneva Convention. Start cleaning now, get the hell out of my platoon," Sergeant Johnson ordered.

The next morning I was saying my goodbyes to the platoon, everyone with exception of Chief Baltier. Sergeant Johnson asked, "Aren't you going to say goodbye to the chief?"

"The chief has his orders for Fort Knox. I am going to hunt the chief up, and of course, I will be a civilian and we are going to have a hell of a good time!" I blurted out.

The chief that day began the process to 1049 to Germany, where I would never find him. Letters from the platoon to the Kid in Kentucky confirmed the reason I never found him on Fort Knox.

A cook who had been with one of the line batteries and I left the base camp and arrived at the 90th Replacement. A flight was supposed to leave out the next morning, but, the plane, a 707, had blown an engine and it had to be replaced.

On the second day at the 90th Replacement mortars came in. They hit the newbie area, the mortars did not kill or injure anyone; however, several men were trampled to death as there was sheer panic among those men. On the third day, the plane arrived at Binh Hoi.

I was the last of the 217 men to board. I stopped on the steps just before leaving, turned around and took a last look at the country that combat veterans referred to as Hell. I thought of my little girl at the orphanage, no one to replace me. I thought no one to buy the next pair of sandals, no one to feed her.

I thought of so many lives lost in the 5/2, Special Ops, and Special Forces, so many lives lost in the whole theater. Next to me, when I sat down, was a SP4 who in the early years of the war was with the 82nd Airborne, and had attended the Ranger School at Fort Benning after his first tour with the 82nd.

As the plane lifted off, three Viet Cong popped up, one RPG rocket was fired, and two with AK-47s were pouring it on. It was so close one could observe the empty brass coming out of the battle rifles.

After the plane gained enough altitude, the pilot came over the intercom with a little speech. He stated that "Charlie" was saying good-bye with the AK-47's and the RPG. He also announced we would arrive at Ft. Dix, New Jersey in about 24 hours. There was a huge cheer among the troops.

Just the Beginning!

Glossary

50-cal
.50 caliber machine gun

AK-47
Soviet-manufactured semi-automatic and full automatic combat assault rifle

Article 15
Section of the Uniform Military Code of Justice, a non-judicial punishment, which permits commanders to administratively discipline troops without a court-martial

ARVN
Army of the Republic of Vietnam (South Vietnam)

A-Team
Basic ten man team of the U.S. Special Forces, which often led irregular military units not responsible to the Vietnamese military command

Base Camp
A resupply base for field units and a location for headquarters of brigade or division size units, artillery batteries and air fields

Battery
An artillery unit equivalent to a company, including six 105mm or 155mm howitzers or two 8-inch or 175mm self-propelled howitzers

Berm
A narrow shelf, path or ledge typically at the top or bottom of a slope; a mound or wall of earth or sand

Broken Arrow
Term used when an American position was overrun by the enemy, to call in an air strike within that position

Brown Bar
A lieutenant, who wore a single bar of rank; in the field the bar was often brown or black (also – Butter Bar)

Clacker
A small hand-held firing device for a claymore mine

Claymore
An antipersonnel mine carried by the infantry, which, when detonated, propelled small steel cubes in a 60-degree fan-shaped pattern to a maximum distance of 100 meters

Connex Container
Corrugated metal packing crate approximately six feet in length

Cowboy
Slang term for a Vietnamese local

C-rations
Combat rations – canned meals for use in the field

Duece and a Half
M35 2 ½ ton cargo truck

Duster
M42 40 mm Self-Propelled Anti-Aircraft Gun that resembled a tank, built for the US Army from 1945 until Dec. 1960

Elephant Grass
Tall, razor-edged tropical plant indigenous to certain parts of Vietnam

Firebase
A secured site from which field artillery can lay down heavy firepower in support of other military units

Greenline
The outer perimeter of a base, so greenline duty was guard duty of the base perimeter

Guard Mount
Official formation where active duty or security personnel line up for inspection and instruction prior to a shift or engagement

Hooch
A hut or simple dwelling, either military or civilian (also spelled "hootch")

Huey
Nickname for the UH-1 series helicopters

I Corps
The northernmost military region in South Vietnam

II Corps
The Central Highlands military region in South Vietnam

III Corps
The densely populated, fertile military region between Saigon and the highlands

IV Corps
The marshy Mekong Delta southernmost military region

Lima-lima
Land line telephone communication network between two points on the ground

M16
The standard U.S. military rifle used in Vietnam beginning in 1966

Napalm
A jellied petroleum substance that burns fiercely and was used as a weapon against enemy personnel

NCO
Non-commissioned officer, typically those with a rank of Sergeant or Corporal

NVA
North Vietnamese Army or People's Army of Vietnam

Profile
A prohibition from certain types of military duty due to injury or disability

Quads / Quad 50s
Weapon consisting of four heavy barrel .50 caliber M2 browning machine guns

Piasters
Vietnamese currency

Puff the Magic Dragon
Douglas AC-47 gunship used for close air support

Sapper
A Viet Cong or NVA commando, usually armed with explosives

Satchel Charge
Pack used by the enemy containing explosives that is dropped or thrown and is generally more powerful than a grenade

Searchlights
Giant, high-powered lights used as spotlights at American base camps during the war

Shaped Charge
An explosive charge focused in one direction

Spec-4
Specialist 4th Class, a rank immediately above Private First Class

Spec-5
Specialist 5th Class, equivalent to a Sergeant

Steel Pot
The standard U.S. Army helmet; the outer metal cover

Tiger Rangers
Vietnamese Rangers of the Army of the Republic of Vietnam, who were trained and assisted by American Special Forces and Ranger advisors

Tops Pops Corn Machine
Slang term for the M55 .50 caliber air defense gun (see Quads)

Trigger Bar
Metal lever that when pulled dischargers a gun

UH-1H
A Huey helicopter

Viet Cong / VC
Political organization and army (Communist-led forces) that operated in South Vietnam that was allied with North Vietnam against the armies of South Vietnam and the United States

Web Gear
Canvas belt and shoulder straps for packing equipment and ammunition on infantry operations

Wood Line
A row of trees at the edge of a field or rice paddy

About David J. McCormack

David J. McCormack was born April 27, 1948 in an apartment above what was for years Leonard's Grocery, about a mile from what is the I-65 interchange on US 62 Elizabethtown, Kentucky. He was born to Blumer Bryan McCormack and his mother, Catherine Virginia Clark McCormack. Post war he worked as an auto mechanic and taught auto mechanics at Elizabethtown Technical College. He opened Mack's Service Center in 1977.

About the Author

Karen Angelucci, Kentucky author and Realtor, lives in Lexington with her two daughters, Sarah and Rachel. Contact Karen at karenangelucci.com or karenangelucci67@gmail.com.

Index

E

F

G

H

I

J

K

L

M

RANGER

DE OPPRESSO LIBER

2
SEAL

Missing in Action:

2,338